Bridging the Gap between Theory and Practice in Educational Research

Bridging the Gap between Theory and Practice in Educational Research

Methods at the Margins

Edited by

Rachelle Winkle-Wagner
Cheryl A. Hunter
Debora Hinderliter Ortloff

BRIDGING THE GAP BETWEEN THEORY AND PRACTICE IN EDUCATIONAL RESEARCH

First published in hardcover in 2009 by PALGRAVE MACMILLAN® in the United States—a division of St. Martin's Press LLC, 175 Fifth Avenue, New York, NY 10010.

Where this book is distributed in the UK, Europe and the rest of the world, this is by Palgrave Macmillan, a division of Macmillan Publishers Limited, registered in England, company number 785998, of Houndmills, Basingstoke, Hampshire RG21 6XS.

Palgrave Macmillan is the global academic imprint of the above companies and has companies and representatives throughout the world.

Palgrave® and Macmillan® are registered trademarks in the United States, the United Kingdom, Europe and other countries.

ISBN: 978–1–137–33826–6

The Library of Congress has cataloged the hardcover edition as follows:

Bridging the gap between theory and practice in educational research : methods at the margins / [edited by] Rachelle Winkle-Wagner, Cheryl A. Hunter, Debora Hinderliter Ortloff.
    p. cm.
Includes bibliographical references and index.
ISBN 0–230–61072–2
    1. Education—Research—United States—Methodology.
2. Minorities—Education—United States. 3. Indians of North America—Education—History. I. Winkle-Wagner, Rachelle. II. Hunter, Cheryl A. III. Ortloff, Debora Hinderliter.

LB1028.25.U6B75 2009
370.7′2—dc22   2008054690

A catalogue record of the book is available from the British Library.

Design by Newgen Imaging Systems (P) Ltd., Chennai, India.

First PALGRAVE MACMILLAN paperback edition: April 2013

10 9 8 7 6 5 4 3 2 1

Contents

Acknowledgments

There are three individuals who deserve to be immensely acknowledged for the creation of this book: Barbara Dennis, Adrea Lawrence, and Joshua Hunter. *Methods at the Margins* was born out of intense conversations with these individuals and a mentoring process that has left a mark on us all. Barbara Dennis was a faculty mentor who challenged and inspired us to follow what was important and to persevere when odds were against us. As each of us graduated, she continued to be a source of inspiration and remained committed to supporting this manuscript and our larger academic goals. She is also a dear friend whom all of us cherish. Each of us has a deep sense of gratitude toward Barbara Dennis for challenging us with in-depth theoretical works, and also appreciation for the sense of trust reposed in us to explore these works without feelings of censure or inadequacy. All of us are of the view that we have grown as scholars because of her insightful mentoring and unwavering commitment to our personal and professional growth. Adrea Lawrence made a commitment to work on this manuscript because her sense of critical reflection upon social injustice is unwavering. She was staunchly dedicated, just as she has been in her own work, to articulating the pervasiveness of marginalization through critical inquiry. She challenges traditional theories and concepts or linear ways of thinking in her own work and did the same for this work too, offering us feedback to hone our analysis and writing. Special thanks must go to Joshua Hunter, who at times would leave our conversations to enjoy the falling snow or the wind blowing through the trees. When you read his work you will understand why. However, his insight in giving feedback and in reviewing, as well as his writing, hopefully brings that much-needed reminder of falling into anthrocentrism and reminds us that if we miss the commitment to the site of our studies, indeed we have lost the commitment to the participants and other lives that exist in those places.

We also want to thank several individuals whose work on the manuscript helped make this book a reality. First, thanks are owed to Julia Cohen and Samantha Hasey at Palgrave Macmillan for their remarkable professionalism and dedication to their craft. In addition, three anonymous reviewers helped make this work sharper and more finely tuned. Finally, we would like to thank the Center for Urban and Multicultural Education (CUME) at Indiana University, under the direction of Joshua Smith, which provided immeasurable support for the completion of this book. In particular, several research assistants, Kristie Coker, Ashley Rittenhouse, Erika Klosterhoff, Jessica Mungro, and CUME's project manager, Shanna Stuckey, tirelessly proofread and formatted the manuscript. Their work, we are sure, was at times tedious, but they pursued it with attention to detail and a sunny disposition. We are grateful for their assistance.

As a group, we made a commitment to take the idea of addressing marginality from a methodological stance and put forward a product that could serve as a springboard for a continued conversation about the ways in which we marginalize people, places, and thoughts. Looking back, we must have realized that we needed something to keep us connected when we all went our disparate ways. Thus, in many ways, this project has solidified bonds that go beyond the academy and into our own lives as partners, spouses, mothers, fathers, colleagues, and, most importantly, friends. The process of writing this book has strengthened us not only as scholars but also as people who want to see a difference in the world we hand over to our children. It is invaluable to have such a wonderful group of friends who believe equally passionately about that endeavor. We have been honored to be part of this group of individuals, and hopefully this work will represent all of our hopes and goals and will continue to ignite our own passions.

Contributors

Michael W. Apple is the John Bascom Professor of Curriculum and Instruction and Educational Policy Studies at the University of Wisconsin, Madison. He has written extensively on the relationships among education, power, and inequality. His most recent books include *The Subaltern Speak: Curriculum, Power, and Educational Struggles* (2006), and *Democratic Schools: Lessons in Powerful Education* (2007).

Phil Francis Carspecken is Professor of Educational Inquiry and Philosophy of Inquiry in the School of Education at Indiana University in Bloomington, Indiana. His research interests include critical ethnography, spirituality, and social science research and critical theory.

Ginette Delandshere is a Professor in the School of Education at Indiana University in Bloomington, Indiana. She directs the Inquiry Methodology Graduate Program and teaches courses in research methodology, measurement and assessment, and statistics applied to social science research. Her research interests include study of the research methodology in social science and study of the assumptions underlying measurement and assessment theories and practices.

Barbara Dennis is an Associate Professor of Inquiry at Indiana University and is concerned about gender and racial inequities and their intersection with schooling. She is mother of two children and an aspiring novelist.

Joshua Hunter is currently writing his dissertation concerning environmental education and sense of place. He keeps busy on the side hiking with his partner, two darling girls, and providing environmental education at an elementary school in Cleveland, Ohio.

Cheryl A. Hunter is an Assistant Professor in the Department of Education at the University of North Dakota. Her research focuses upon the intersections of gender and education. She also pursues her interests in cultural immersion, sociology of education, and is a qualitative research methodologist.

Adrea Lawrence is an Assistant Professor in the School of Education, Teaching, and Health at American University in Washington, DC. Her research focuses on American Indian education history and disciplinary learning within the social studies.

Amaury Nora is a Professor of Higher Education, Director of the Center for Student Success in the College of Education at the University of Houston, Houston, Texas, and Editor of *The Review of Higher Education*, the journal of the Association for the Study of Higher Education (ASHE). His research focuses on college student persistence, the adjustment to and engagement in colleges with diverse student populations across different types of institutions, the development of retention models that integrate disciplinary theoretical perspectives with college persistence theories, graduate education, and theory building and testing.

Debora Hinderliter Ortloff is an Assistant Professor at the University of Houston Clear Lake. Her research interests center around issues of global citizenship, multicultural education reform, and research methodology. She is currently principal investigator on a statewide examination of international education in Indiana.

Elizabethe C. Payne is an Education Sociologist in the Department of the Cultural Foundations of Education at Syracuse University. She is also the founding director of The Q Center @ ACR, a youth center supporting Lesbian, Gay, Bisexual, and Transgendered (LGBT) young people in the Central New York area. Elizabethe's research focuses on the life histories of adolescent lesbians, school experiences of LGBT youth, and HIV and sex education. She teaches courses in qualitative research methodology, youth culture, and queer youth experiences in schools.

Donald Warren is Professor Emeritus, History of Education and Policy, and University Dean Emeritus at the Indiana University School of Education. His recent publications focus on civic learning in U.S. history, slavery as an educational institution, and American Indian history as an educational narrative. All are related to his long-standing interests in education historiography.

Rachelle Winkle-Wagner is an Assistant Professor of Higher Education at the University of Wisconsin-Madison. Her research focuses on the sociological aspects of race, class, and gender in higher education, and it considers topics such as access and retention for students of color, race and identity theory, the application of philosophy to research, and qualitative inquiry. She is the author of the book, *The Unchosen Me: Race, Gender, and Identity among Black Women in college* (The Johns Hopkins University Press).

Skye Winter, the cover artist, was born and raised in the U.S. Virgin Islands. As daughter of the renowned Caribbean artist, Eric Winter, Skye developed a unique eye for beauty. Now residing in South Beach, Florida, Skye expresses her talents by creating stunning photographic images of the serene as well as the vibrant, and also the bold and surrealist, art with her paintbrush.

Tarajean Yazzie-Mintz (Navajo), an Assistant Professor in Curriculum and Instruction at Indiana University, is interested in documenting teacher knowledge, particularly how teachers theorize and conceptualize pedagogy. Her work contributes to educational research in the areas of cross-cultural research, American Indian education, cultural/language learning, and educational practices in and outside schools.

Aki Yonehara is a postdoctoral research fellow at Japan Society for the Promotion of Science. Her academic interests are in educational development policy in developing countries, theoretical consideration on human development, and quantitative social research methods. She has worked as an educational consultant on international developmental projects of the Japan International Cooperation Agency (JICA).

INTRODUCTION

The Not-Center? The Margins and Educational Research

Rachelle Winkle-Wagner, Debora Hinderliter
Ortloff, and Cheryl A. Hunter

> To be in the margins is to be part of the whole but outside the main body.
> —bell hooks, 2000

We are not the first to highlight marginality and the margins. Scholars, poets, artists, and everyday people have lived in the margins for centuries. In the United States, entire populations of people have been pushed from the center, either physically or psychologically, toward the sides—the margins—of society. Likewise, research about these groups, and research that is largely excluded from the center for one reason or another, is often pushed or pulled from the center over to the margins because it is essentially excluded from the center by publishing practices, academic traditions, and discussions about research methodologies. To be at the margins in many ways means that one or one's ideas/ traditions/methods/personhood is not-center, external to that which is mainstream.

bell hooks (2000) reflects on her own experience of marginality, of being born into the margins as a black woman in the United States:

For black Americans living in a small Kentucky town, the railroad tracks were a daily reminder of our marginality. Across those tracks were paved streets, stores we could not enter, restaurants we could not eat in, and people we could not look directly in the face. Across those tracks was a world we could work in as maids, as janitors, as prostitutes, as long as it was in a service capacity. We could enter the world, but we could not live there. We had always to return to the margin, to beyond the tracks, to shacks and abandoned warehouses on the edge of town (p. xvi).

It is this notion, of being able to enter but not exist in the center, that this book takes up in its focus on the margins and marginality in research processes. This volume explores that research which attempts to enter the center, but, which may not really exist there. This book presents research about marginalized groups and research theories or techniques that are marginalized, pushed away, and thereby commonly excluded.

This book is at once a practical, theoretical, and methodological exploration of the margins and marginality in educational research. It offers theories, research methods or techniques, and exemplars of full research studies, all with the common goal of understanding, exploring, and shedding light on marginality and the margins in research processes. This volume attempts to bridge the gaps between theory and practice[1] in research while simultaneously providing a conduit to the divide between qualitative and quantitative research. This does not mean that the book provides "mixed methods" research. Rather, it provides an investigation of the margins from both qualitative and quantitative traditions. At times, these methods are mixed or complementary within a single study, but generally they are explored in their own right in this volume.

Methods at the Margins provides a practical contribution offering new and/or alternate ways of thinking about educational processes and research about these processes. Methodologically, this book provides concrete examples of research techniques for those conducting research with underrepresented, marginalized populations or about marginalized ideas. Theoretically, this project asserts new theoretical models as they relate to research methods, knowledge more generally, and the study of underrepresented groups. Finally, this volume offers exemplars of educational research at the margins, providing instances of innovative or alternative methodologies, paradigms, theoretical frameworks, or ways to present findings. Education and educational research often extend beyond schools and classrooms. In this way, educational research is instructive in a very broad sense in ways that other fields are not.

Through an examination of methods at the margins, the authors refer not only to research that works with those in marginalized groups but also point to the process through which some ideas, groups, and innovations are prevented from being centered and are kept at the margins. Ultimately, this volume aims at expanding knowledge itself—altering the center by allowing the margins to inform it—allowing knowledge to be created and extended to include those ways of knowing that have historically been unexplored or ignored.

The Margins as the Not-Center? Redefining a Deficit Model

The margins can be defined by that which lies outside of the boundaries—be they sociohistorical, theoretical, methodological, or epistemological. The margins are the places at which ideas, practices, or persons become a challenge to the dominant culture. The center or mainstream often defines the margins through boundaries that put up limits, define, or exclude. Through the center, one can exclude certain ideas, groups, or ways of being—even if unwittingly. Yet, to provide a working definition of the margins by the boundaries of dominant culture is to define those in the margins as *not* something. Marginality becomes defined as the not-normal, the not-mainstream, the not-center; invoking a deficit model upon those people, ideas, and so on that are marginalized while reaffirming the normalcy of the center.

Likewise, to define the margins too clearly may result in a bounding of that which was once innovative, making it central or part of the mainstream, the whole. Yet, to leave the margins undefined and unexamined may be to miss out on valuable insight or knowledge that is unable to be viewed from a position in the center. The authors' primary task in this book is to shed light on this paradox, this seemingly undefinable concept, of margins and marginality and the way margins and the people who are marginalized are a part of research processes.

While margins seemingly refer to a place or a space, even if that space is outside or is limited, marginality is an adjective to describe one's location within that place or space. Marginality inherently references the mainstream. That is, to be at or in the margins as an individual or a group is to at once have an understanding of the mainstream while also knowing that one is not a commonly accepted part of it. In individual terms, through marginality, a limitation is placed upon the individual's ability to self-actualize his or her identity (Cornell, 1998). In the same way that limitations on a person's identity are a result of marginality, innovations and ideas are curtailed when marginalization is allowed to stand.

Throughout history, many theorists and scholars from racial minority groups have described this as a sense of two-ness. For example, W. E. B. DuBois (1903) described being African American in the United

States as a necessary dual consciousness, a two-ness, an understanding of the majority White groups in addition to the minority Black groups in the United States. While never being able to gain full membership in the majority White group, his minority status resulted in him experiencing a two-ness or needing to understand both groups simultaneously, even when remaining in his marginality. Nearly a century later, bell hooks (2000) corroborated this notion of two-ness in her reflection on growing up Black in the U.S. south:

> There were laws to ensure our return. To not return was to risk being punished. Living as we did—on the edge—we developed a particular way of seeing reality. We looked both from the outside in and from the inside out. We focused our attention on the center as well as on the margin. We understood both. This mode of seeing reminded us of the existence of a whole universe, a main body made up of both margin and center. Our survival depended on an ongoing public awareness of the separation between margin and center and an ongoing private acknowledgement that we were a necessary, vital part of that whole (p. xvi).

It is this assertion that those ideas and peoples on the margins are a vital part of the whole that provides the impetus for this book. The margins connote both-ness, two-ness, the ability to simultaneously understand that which is central and that which is outside of the center, the marginal. While those on the margins, or in the minority, must at once understand both the majority/mainstream and the minority/margin, those in the mainstream typically only understand the mainstream. It is for this reason that a description of the margins could alter the center or the mainstream—it could change what is known about education, society, and the larger world.

Perhaps the margins could actually become the center, although not in a hegemonic way. That is to say, the margins could redefine the center in a more inclusive, holistic, altruistic way. Nobel Prize winning author Elie Wiesel (1986), in his memory of the Holocaust, eloquently initiates this idea:

> We must take sides. Neutrality helps the oppressor, never the victim. Silence encourages the tormentor, never the tormented. Sometimes we must interfere. When human lives are endangered, when human dignity is in jeopardy, national borders and sensitivities become irrelevant. Wherever men and women are persecuted because of their race, religion, or political views, that place must—at that moment—become the center of the universe.

Here, Wiesel suggests that in the name of social justice and in an effort to fight oppression, the margins must become the "center of the universe." This idea of using the margins to transform the center is the goal of this book. Our goal is to enlighten the mainstream to the margins while not losing the innovation and creativity of the margins. To shed light on that which is outside of the whole or the center may actually deepen and strengthen the center by offering new ways of knowing about the world.

The definition of the margins is purposefully flexible in this book, allowing each contributing authors to grapple with these issues for themselves. That said, discussing the margins does connote some sense of boundaries that are still in place at some level. The margins are a form of subtext in this way. Thus, while our formal definition of marginality and the margins remains somewhat flexible, it is a bounded flexibility, and all innovations or all things outside of the center are not included in our discussion of the margins. Likewise, all possible innovations in research cannot be discussed in a single volume. We attempt to tackle the subject while simultaneously knowing that the development of this volume will itself begin to assert a definition of margins and marginality that may be different from others, or may even differ from the definition offered here in this introductory chapter.

Understanding the Margins Theoretically

Philosophers and theorists have explored marginality in many ways. The theoretical portion of this book will continue this theoretical dialogue on margins and marginality. As an example of philosophical notions of marginality, Hegel developed a master/slave dichotomy to explicate the interaction between an oppressor and the oppressed, between those in the mainstream and those on the margins. In this dichotomy, Hegel (1807/1977), similar to DuBois' (1903) double consciousness, described the way in which a slave must simultaneously know the master and his/her own self, while the master need only know the domain (i.e., needs, desires, etc.) of the master. In this case, to be at the margins is to know more than those in the mainstream. In this book, this Hegelian and DuBoisian notion of two-ness leads us to explore research that focuses on minority or marginalized groups as well as to interrogate ways in which the research process can move away from reinforcing and reproducing marginalization. As a way to understand the wisdom and insight of those who experience two-ness, many of the authors describe research of, for, with, and about groups, knowledge, techniques, or ideas often underrepresented or unrecognized within the mainstream.

Scholars in the philosophical tradition of critical theory often reveal the margins. This multidisciplinary approach to studying society maintained a keen interest in emancipation from oppression and a commitment to freedom, happiness, and a rational ordering of society. This tradition of theory is intrinsically open to reflection, revision, and development in a search for emancipation from the existing social order. Critical educational research refers to research that has as its goal the fight against oppression and uncovering societal inequalities, attempting to push for social change within educational processes (Kincheloe and McLaren, 2003; Carspecken, 1996; McLaren and Giarelli, 1995). Rather than be limited to one particular way of conducting research, it includes methodologies that facilitate the examination of power disparities within social systems and groups.

This book is certainly critical in nature, although the label "critical research" is both broader and narrower than the content of this book. The socio-critical values of critical research are central to the work of this text in that this project is primarily concerned with the role of educational research in emancipating individuals and society from those aspects of the social order that are oppressive, which are ideas that are drawn from the critical tradition. This book is narrower in scope because it focuses specifically on innovative methods within the practice of critical research. It is broader in scope because the authors may or may not agree entirely with current critical theory—in the theoretical development in this volume, the authors in many ways work to expand the notion of critical theory. In addition, this book includes quantitative approaches. Usually, studies that carry a reference to criticalism are qualitative. This book hopes to introduce quantitative studies into the field of critical educational research. This requires extending critical methodological theory to more explicitly explain quantitative methods. In this way, this volume redefines criticalism in research.

The Margins and Educational Research

In educational research, to be at the margins means that one's group of study, one's theoretical position or epistemology, or one's research technique is outside of mainstream research—outside of the center. Similar to the way that a marginalized person or a person from a minority group could simultaneously understand both the mainstream and the margins/minority, those doing marginalized research have a unique perspective.

They must understand the center, the mainstream of research processes, and that which is excluded from the center.

Research in general is the process of creating knowledge, of attempting to answer questions, create theory, and grapple with previously held theories. What is more, research has as its purpose the uncovering of that which has been previously covered—both those things that are controversial and those things that fit into mainstream ways of knowing. *Methods at the Margins* is a compilation of research questions, theories, and exemplars that address the most central questions in the research process: Who does the research process represent and who are excluded? How does one research and understand educational experiences, systems, and related phenomena in potent ways? In what ways does the tradition of academic research reinforce the status quo, protecting the center and working against the innovations that might come from the margins? While mainstream research arguably may represent the experiences of the majority populations, this edited volume is concerned primarily with those who have historically not been represented in the research process, including the writing, or reporting of research. That is to say, this book takes as its primary task the voicing of those theories, research practices, persons, and data that could uncover serious limitations in mainstream research, societal, and educational issues that have not been discussed, and also attempts to present research in unexpected ways, providing new forms of knowledge. Using methods on the margins and looking at marginal cases are potent because they disrupt what is expected.

This book is organized into three parts. Each part commences with an introduction that weaves together the methodological arguments that emerge among the chapters. Part I provides a theoretical exploration of objectivity, scientism, and "scientifically-based research," as the preponderance of these perspectives marginalize research that does not provide evidence based on natural sciences' definitions of objectivity, regardless of whether or not that research provides a legitimate accounting of knowledge in a community. Michael Apple leads off with an argument that scientifically-based research and the environment it creates for education researchers has a potential effect upon education and knowledge creation more generally. Apple provides a clarion call for those doing critical research—offering concrete suggestions about tasks for those scholars or practitioners engaging in this work. Building on this call for change, Ginette Delandshere reflects on the recent political climate and the trend toward scientific, evidence-based research. She describes the impact of this political climate on research methodologies and on the evidence that comes from the dominance of these methods. Phil Carspecken's chapter considers the theoretical limits of knowledge more generally, exploring the way these limits may eventually restrict the self or one's identity. Carspecken's chapter provides the theoretical potential for methodological and knowledge-creating innovations. In Barbara Dennis's chapter, she provides a theoretically based rationale and a charge for conducting research at the margins, laying the groundwork for research that liberates those reading, conducting, or agreeing to be involved in research processes. The first part of this book thus frames a meta-theory, knitting together many of the creative endeavors to follow.

In Part II, a variety of methodological techniques, examples, and research designs are asserted. This part of the book provides practical guidance for conducting a variety of studies using marginal methodologies. The methods themselves are not meant to be codified, but they instead point toward the way methodology and substance are recursively and innovatively interesting. Each chapter in this part will lodge a concurrent critique of traditional educational research practice. One of the points illustrated is how methods and substantive fields are integrated at the margins where research is guided by reform. Thus, each of these chapters brings the margins forward methodologically and demonstrates how these margins simultaneously bring forward the potential to unsettle inequity, challenge ideology, and offer hope for socio-critical transformation. In an examination of learning through "place" in the research process, Joshua Hunter provides in his chapter an exemplar of research methodology that does not focus on people-over-nature in the way that mainstream work often does in its use of nonanthropocentric deep ecology. Don Warren turns mainstream historical work and definitions of education on their proverbial heads through his nonchronological, historical consideration of American Indian initiated history and education, examining this history as a series of episodes in order to provide a clearer look into indigenous societies before they came into contact with European settlers. Amaury Nora provides a charge for moving away from "one-model-fits-all" approaches in the study of Hispanic college students, calling for a bridge between qualitative and quantitative traditions. Finally, in chapter 8, Winkle-Wagner considers the issue of validation in cross-racial research, providing insight from her own work with African American women as a guide for critically oriented validation techniques. All of these methodological offerings provide examples of methodological techniques or ideas that are often marginalized or are at the margins of mainstream research.

The final part, Part III, of the book presents exemplars of research conducted at the margins. The exemplars, like the theories and practical methodological techniques, are opportunities to see the innovations unfolding specifically and to examine the findings in terms of theory, practice, and socio-critical value. These examples are not meant to be exhaustive, and are rather illustrative. Adrea Lawrence provides an example of a nonchronological, microhistorical approach to historical writing that innovatively combines qualitative and historical analysis in her work with letters written by a White teacher of American Indian students in the early 1900s. Through Portraiture methodology, the chapter by Tarajean Yazzie-Mintz provides a model of purposeful and meaningful research in American Indian education that is rooted in Native-defined purpose and needs rather than simply those of the researcher. The chapter by Elizabethe Payne examines the gender-related identity construction of lesbian adolescents, offering an exemplar of the study of agency and resistance through critical life story methodology. Cheryl Hunter's ethnographic work with adolescent teens in a sexuality education course proffers a pioneering example of a traditional methodology written in a nontraditional format. In her exemplar of substantive theoretical research using Germany as a country context, Debora Hinderliter Ortloff delves into the way that citizenship education can contribute to marginalization, affecting the development of education policy. Finally, Aki Yonehara provides a critique of, and an alternative to, mainstream statistical modeling in her work with basic education in Tanzania, presenting a theoretical and practical exemplar for the use of Hierarchical Generalized Linear Modeling (HGLM).

In investigating marginality, this book cannot claim to include all instances and types of marginal research methods, practices, and findings. Hopefully these chapters will serve as a springboard for future books dedicated to more marginalized voices that could not be included in this work. We hope that the critiques asserted here act as an impetus toward more writing on marginality, for this work is not the culmination of a project that addresses marginality but, rather, the birth of one.

The margins connote a dialectical tension between the part and the whole, between the universal and the particular, and between the mainstream majority and the peripheral minority. To invoke the margins is to shed light on what lies at the edge, the border, the fringe, the periphery, the limit, or the excluded. One may initially consider all that lies at the margins to be quirky, odd, strange, or abnormal. In addition to being inherently outside of the center or the mainstream, the margins may summon innovation, creativity, eccentricity, nonconformity, and newness. The margins may lead to revolution, change, and to a redefining of what is known. This is the interest of this book—to contemplate change and newness in educational research.

The purpose of this volume is to introduce theory, techniques, and exemplars that deal with issues of marginality, considering, but not limited to, such questions as: What does it mean to be marginal? Are all people and/or research methodologies marginal in some senses and mainstream in others? Is it correct to define the margin as anti-mainstream? Are these issues all that have to do with margins, or does a margin have other characteristics? Are all margins equal? Do some margins have more potential for creativity than others? Why is it so, or why is it not so? Do some centers have more potential than others? Why so or why not so? Do methodological margins always exist, do they always offer the same possibilities, or is the nature (and potential) of a margin dependent on the center that makes its existence possible? The chapters of this book provide various and sometimes debatable answers to these questions.

The Beginning

This project was initiated from a collective process involving a conversation among friends about our own works and the way we felt marginalized because of our methods, the populations/topics we were studying, and our desire to report on and do our work in innovative ways. The people who were initially involved in this conversation—Barbara Dennis, Adrea Lawrence, Joshua Hunter, Debora Hinderliter Ortoff, Cheryl Hunter, and Rachelle Winkle-Wagner—provided a collective space for innovation and transformation. It is our hope that this book also offers such a space for research to those studying it, reading it, and conducting research based on it to transform those involved and society at large.

Building on this initial conversation, these chapters provide a bridge between qualitative and quantitative methods and between theory and practice. All of the chapters in this volume—be they theoretical, methodological, or exemplars—have at their heart a connection with those involved in the research. That is, the research, theories, and methodologies presented here are not simply about a particular population or issue; they are also for and related to those persons and issues being studied. There are implications within these chapters for ethics in research, the building and maintenance of trust with research participants, collecting and analyzing data, the validation of research, the creation of new knowledge, and methodological rigor. There are also ramifications in these chapters for giving research back to the communities from which such research stems—using research to affect change and social justice. Embedded both explicitly and implicitly in these chapters is a critique of mainstream, status quo research. While carrying out this critique, the authors simultaneously offer alternatives that hope to liberate, emancipate, and transform those involved in research, those being researched, and larger social forces. What follows is a journey out toward the methodological margins—a journey in an effort to transform the center of educational research and, arguably, society.

Note

1. Practice, for the purposes of this discussion, refers to the action of doing research or the action of working with marginalized groups or issues. While some authors in the book only implicitly refer to practice, there are implications for the practice of research within every chapter.

PART I

Theory

The Politics of Knowledge: What an Examination of the Margins Can Offer

Barbara Dennis

I remember my first high school biology class. My story will sound familiar to almost anyone who attended a U.S. public high school in the sixties or seventies. As a part of learning the scientific method, it was impressed upon us that we must make observations, identify hypotheses, organize methods that objectively test those hypotheses, and then accept the findings. It was difficult for me to challenge the expectations of this science in any formal way, but I experienced the preformation of challenges in an existential way. Biology lab was my induction into what I was led to believe was the *real* work of science. That *real* work of science required us to experiment on THINGS in a sterile, controlled, laboratory setting. From examining our own blood to watching the movement of one-celled organisms to exploring the complex systems of worms, fruit flies, frogs, and mud puppies. In each case, I was taught to examine the object of study from an aloof perspective, as if I was analyzing the elements of a map. To do this, I had to deny something. I had to deny feelings associated with separating myself from the life under the microscope. I had to deny that there was any subjectivity at all to the objects of study merely because humans COULD put them under a microscope and because science's answers held a promise expected to exceed any loss of life or substance. I had to deny that I knew things unscientifically that I was actually quite certain were not opinion. I had to deny the aporia of a method that claimed it was the ONLY route to legitimate knowledge and was yet unable to explain itself through its own means. I had to accept that my own identity as a student of science required these and other denials.

I, along with many other young students, developed the idea that "science" was the route to accurate knowledge. Coupled with this was the attention my education paid to making a strong distinction between facts and opinion, while simultaneously privileging the fact side of the dichotomy as the *real* knowledge. These lessons conspired to make it seem *as if* science and its imitators held the trophy on truth. Objectivity was the golden criterion for this truth. If Michael Apple had asked me, at that time, what counted as research? I would have answered, "properly performed objective science." I had successfully accepted the wisdom of the day: The natural sciences stood at the pinnacle of wisdom in substance and method.

What does this have to do with a book on educational research methods? How does the scientific method as I learned it in high school relate to the contemporary study of human social life or educational practices in general or at the margins? To begin with, this "official" knowledge continues its work in the contemporary educational research community through the call for scientifically based "randomized controlled trials." The hegemony of objectivity legitimizes the hierarchical arrangements of "research" in the education community. Michael Apple provides us with an account of what it can look like to critically engage with the legitimized knowledge of a community. Ginette Delandshere exemplifies this as she calls into question the hegemony of randomized controlled trials in educational research and the policing of schools through evidence-based practices. Phil Carspecken takes us explicitly into a critique of the ideology of scientism, which he defines as the *belief* that ANYTHING counting as knowledge is so because it was obtained through the same procedures employed in the natural sciences and on these same grounds that knowledge is distinguished from opinion and *belief* (notice the inherent contradiction here). These three chapters indicate that my experiences in biology decades ago are connected to the debates and practices of contemporary educational research. They also lodge distinct, but intersecting, critiques of the status quo in educational research. Their critiques map out the margins. The final chapter in this section, which is by me, is really an ethical call suggesting that the margins cannot be dismissed in educational research without simultaneously inviting factual, ethical, and metatheoretical problems.

In my ninth grade biology course I had to deny an interest/concern for anything outside the experiment. My values and ideals were to be left at the laboratory door in very concrete ways. Though some educational researchers advocate this approach, it is not a necessary or perhaps even viable suggestion. You will find that the chapters of this section both independently and collectively propose a provisional, sometimes implicit, ideal. You will also find that the ideal always engages the margins. For Apple, both the motivation and the outcome ideally involve equity, social justice, freedom from oppression, and so on for *all* of *us*—it is an inclusive proposition. Delandshere suggests that knowledge as a whole (that is, our understanding of social life, in general, and educational practices, in particular) is insufficient when it excludes methods, questions, and interests that seemingly lie outside its political definitions and mandates. While Apple's chapter focuses on the possibilities of critical engagement in education, Delandshere identifies the need for a more comprehensive approach to educational research. Apple, Delandshere, and I explicitly argue that our efforts as educators, activists, and researchers must include people, ideas, and methods that might be (at any given time and place) marginalized by the politics of knowing. Carspecken's chapter takes us deeper into the promise of an ideal inclusivity by outlining an expanded view of knowledge, one that might find a place for something traditionally considered the antithesis of knowledge—spirituality. To take a non-metaphysical view of spirituality is to call into question the very constitution of boundaries between knowledge and the self. This effort hints at the metatheoretical possibilities of exploring some of our most taken-for-granted margins in the way knowledge is delineated in contemporary social science in general. My chapter puts the perspectives of Apple, Delandshere, and Carspecken together by fleshing out the necessity for the margins in attempts to make the world a better place. In my biology class it seemed that I had to subvert my concerns to the means-end rationality of natural science.

To succeed in my biology class I had to conceive of knowledge and experimentation through a procedural-object rationality (an emphasis on procedures that mask the subjectivity of the researcher and treat the interest of research as an object). Most people talk about this as a subject-object dualism in natural science, but in my experience I was also expected to deny my own subjectivity. The role of objectivism was confusing because it required me to act objectively both toward the THINGS under my microscope as well as toward my own feelings and experiences in this means-ends view of scientific procedures. Objectivism as a privileged form of knowledge is important to each of the chapters in this section. Most importantly the objectivist view of research is complicated by what is considered a more basic characteristic of research: the subject-subject dialogue. In each of these chapters the principles of dialogue and communicatively achieved understanding are primary, rather than secondary, to the aims and practice of scholarship. Carspecken takes us deeply into the meaning of the subject-subject ground of knowledge (intersubjectivity), while Delandshere explores the implications of a politics that limits the dialogue and intersubjective achievements of educational research by defining its methods in a hegemonic way. Apple extols the possibility of dialogue for engaging criticalists in bringing to fruition the complicated and problematical call for equity, social justice, sustainability, and liberation. My chapter deals with the role dialogue plays factually in the production of research, and, as the other authors here, I argue that this dialogue must be inclusive. This inclusivity is easy to dismiss in a subject-object view of research or in a procedural-object rationality. This is the most important element of all of the chapters in this part. The way in which certainty is anchored to objectivity and its imitators is also called into question. Carspecken's chapter particularly takes us to a new way of thinking about these very basic ideas in social science.

When I was taking biology, I had no tangible choice in the manner or matter of instruction. There was an existing power relation involved in the centrality of science in the curriculum and the requirements of the laboratory course itself. It is quite common, you might think, to deny such choice to high school students, but perhaps these taken-for-granted power arrangements signify something well beyond my mere option to either participate or fail. False choices that cannot be uncovered point toward the politics acting through and behind the backs of actors. Ideology, hegemony, and politics effectively limit both our engagement as subjects with a field of knowledge and practices and our self-reflection and understanding. Delandshere addresses the politics affecting the educational research practices directly. Apple has, throughout his scholarship, drawn our attention to the politics of official knowledge and his chapter in this book describes the possibility for working through those politics toward a more egalitarian and better world.

Michael Apple's chapter continues an ongoing reformist dialogue regarding what a criticalist might do in terms of engaging with the "official" knowledge or hegemony of a community. He proposes that an adequate response to inequity, oppression, and hegemony is one that is mindful of the margins. For Apple, the margins include both the counter-hegemonic knowledge and reformers/activists who work against hegemony for a better world. Ginette Delandshere's chapter is a careful analysis of the scientifically based research politics in the United States. She argues for the practical political inclusion of methods at the margin of this present-day, flawed movement in education. For Delandshere, the margins are those methods/methodologists that remain outside the center of evidence-based educational practices. For both Apple and

Delandshere, the margins are politically and hierarchically constituted. Phil Carspecken's chapter points through the scientism by taking a careful look at the methodological theory of hermeneutics and self-reflection. He doesn't leave us there; he takes us to the brink of spiritual self-discovery, for example. Furthermore, Carspecken locates some additional margins visible in self-reflection. He wrote, "the 'me' part of the self *only* exists through our own position-taking with possible other subject positions, and we find ourselves through this intersubjective reflection as our self-knowing." Each of these authors suggests the benefits of denying the hegemony of the center through the inclusion of the margins. Likewise, each confronts this differently with different ramifications and insights. Taken together, this part offers practical, theoretical, and methodological premises for research work at the margins.

CHAPTER 1

On the Tasks of the Critical Educational Scholar/Activist

Michael W. Apple

Education and Power

Over the past three decades I have been dealing with a number of "simple" questions. I have been deeply concerned about the relationship between culture and power; about the relationship among the economic, political, and cultural spheres (see Apple and Weis, 1983); about the multiple and contradictory dynamics of power and social movements that make education such a site of conflict and struggle; and about what all this means for educational work. In essence, I have been trying to answer a question that was put so clearly in the United States by radical educator George Counts (1932) when he asked, "Dare the School Build a New Social Order?"

Counts was a person of his time, and the ways he both asked and answered this question were a bit naïve. But the tradition of radically interrogating schools and other pedagogic sites; of asking who benefits from their dominant forms of curricula, teaching, evaluation, and policy; of arguing about what they might do differently; and of asking searching questions of what would have to change in order for this to happen are all matters that have worked through me and a considerable number of other people. We stand on the shoulders of many others who have taken such issues seriously, and in a time of neoliberal attacks with their ensuing loss of collective memory, I hope to have contributed to the recovery of the collective memory of this tradition and to pushing it further along conceptually, historically, empirically, and practically. In the process, I have focused much of my attention on formal institutions of schooling and on social movements that influence them.

Of course, no author does this by herself or himself. This is a collective enterprise. And no one who takes these questions seriously can answer them fully or without contradictions or even wrong turns and mistakes. As a collective project, it is one in which we not only stand on the shoulders of those whose work we draw upon critically but also one in which thoughtful criticism of our work is essential to progress. Compelling arguments cannot be built unless they are subjected to the light of others' thoughtful analyses of the strengths and limits of our claims. I want to do some of that self-reflective analysis here. Thus, my arguments are meant to be just as powerful a reminder to me as they are to the reader.

One of the guiding questions within the field of education is a deceptively simple one: "What knowledge is of most worth?" Over the past four decades, an extensive tradition has grown around a restatement of that question. Rather than "*What* knowledge is of most worth?" the question has been reframed. It has become "*Whose* knowledge is of most worth?" (Apple, 2004; 2000; 1996). There are dangers associated with such a move, of course, including impulses toward reductionism and essentialism. These dangers arise when we assume, as some people have, that there is always a one-to-one correspondence between any knowledge that is seen as "legitimate" or "official" and the dominant groups' understanding of the world. This is too simplistic, since official knowledge is often the result of struggles and compromises and, at times, can represent crucial victories, not only defeats, by subaltern groups (Apple, 2000; Apple and Buras, 2006). However, the transformation of the question has led to immense progress in our understanding of the cultural politics of education in general and of the relations among educational policies, curricula, teaching, evaluation, and differential power. Indeed, some of the most significant work on the intimate connections between culture and power have come out of the area of the sociology of school knowledge and critical educational studies in general.

In the process of making the conceptual, historical, and empirical gains associated with this move, there has been an accompanying internationalization of the issues involved. Thus, issues of the cultural assemblages associated with empire and previous and current imperial projects have become more visible. Hence, for example, there has been an increasing recognition that critical educational studies must turn to issues of the global, of the colonial imagination and to postcolonial approaches, in order to come to grips with the complex and, at times, contradictory synchronic and diachronic relations between knowledge and power, between the state and education, and between civil society and the political imaginary.

For example, under the influences of a variety of critical works on the history of literacy and on the politics of popular culture. Raymond Williams's work was crucial historically here (See Williams, 1961; 1977; See also Apple, 2004). As in a number of other fields, it became ever clearer to those of us in education that the very notion of the canon of "official knowledge" had much of its history in a conscious attempt to "civilize" both the working class and the "natives" of an expanding empire (Apple, 2000). The very idea of teaching the "Other" was a significant change. For many years in Europe and Latin America, for example, the fear of working class and "peasant" literacy was very visible. This will be more than a little familiar to those with an interest in the history of the relationship among books, literacy, and popular movements. Books themselves, and one's ability to read them, have been inherently caught up in cultural politics. Take the case of Voltaire, that leader of the Enlightenment who so wanted to become a member of the nobility. For him, the Enlightenment should begin with the "grands." Only when it had captured the hearts and minds of society's commanding heights could it concern itself with the masses below. But, for Voltaire and many of his followers, one caution should be taken very seriously; one should take care to prevent the masses from learning to read (Darnton, 1982). This, of course, was reinscribed in often murderous ways in the prohibitions against teaching enslaved peoples how to read (although there is new historical evidence that documents that many enslaved people who were brought to the Americas were Muslim and may already have been literate in Arabic).

Such changes in how education and literacy were thought about did not simply happen accidentally. They were (and are) the results of struggles over who has the right to be called a person, over what it means to be educated, over what counts as official or legitimate knowledge, and over who has the authority to speak to these issues (Apple, 2000; Mills, 1997). These struggles need to be thought about using a range of critical tools, among them analyses based on theories of the state, of globalization, of the postcolonial, and so much more. But none of this is or will be easy. In fact, our work may be filled with contradictions. Take for instance the recent (and largely justifiable) attention being given to issues of globalization and postcolonialism in critical education, to which I turn in my next section.

Globalization, Postcolonialism, and Education

At the outset, let me be honest. I no longer have any idea what the words "globalization" and "postcolonial" mean. They have become sliding signifiers, concepts with such a multiplicity of meanings that their actual meaning in any given context can only be determined by their use. As Wittgenstein (1963) and others remind us, language can be employed to do an impressive array of things. It can be used to describe, illuminate, control, legitimate, mobilize, and many other things. The language of postcolonialism(s) (the plural is important), for example, has many uses. However, all too often it has become something of a "ceremonial slogan," a word that is publicly offered so that the reader may recognize that the author is *au courant* in the latest linguistic forms. Its employment by an author here is largely part of the conversion strategies so well captured by Bourdieu in *Distinction* (1984) and *Homo Academicus* (1988). Linguistic and cultural capitals are performed publicly to gain mobility within the social field of the academy. In my most cynical moments, I worry that this is at times all too dominant within the largely white academy.

But, of course, the postcolonial experience(s) (and again the plural is important) and the theories of globalization that have been dialectically related to them are also powerful ways of critically engaging with the politics of empire and with the ways in which culture, economy, and politics all interact globally and locally in complex and overdetermined ways. Indeed, the very notions of postcolonialism and globalization "can be thought of as a site of dialogic encounter that pushes us to examine center/periphery relations and conditions with specificity, wherever we may find them" (Dimitriadis and McCarthy, 2001, 7).

As they have influenced critical educational efforts, some of the core politics behind postcolonial positions are summarized well by Dimitriadis and McCarthy (2001) when they state that "The work of the postcolonial imagination subverts extant power relations, questions authority, and destabilizes received traditions of identity" (10; see also Bhabha, 1994 and Spivak, 1988).

Educators interested in globalization, in neoliberal depredations, and in postcolonial positions have largely taken them to mean the following. They imply a conscious process of repositioning, of "turning the world upside down" (Young, 2003, 2). They mean that the world is seen relationally—as being made up of relations of dominance and subordination and of movements, cultures, and identities that seek to interrupt these relations. They also mean that if you are someone who has

been excluded by the "west's" dominant voices geographically, econom-
ically, politically, and/or culturally, or you are inside the west but not
really part of it, then "postcolonialism offers you a way of seeing things
differently, a language and a politics in which your interests come first,
not last" (2). Some of the best work in the field of education mirrors
Robert Young's (2003) more general claim that postcolonialism and the
global sensitivities that accompany it speak to a politics and a "philosophy of activism" that involve contesting these disparities. It extends the
anticolonial struggles that have such a long history and asserts ways of
acting that challenge "western" ways of interpreting the world (4). This
is best stated by Young (2003) in the following two quotes:

> Above all, postcolonialism seeks to intervene, to force its alternative knowl-
> edges into the power structures of the west as well as the non-west. It seeks
> to change the way people think, the way they behave, to produce a more just
> and equitable relation between different people of the world. (7)
>   and Postcolonialism... is a general name for those insurgent knowl-
> edges that come from the subaltern, the dispossessed, and seek to change
> the terms under which we all live. (20)

Of course, what Young says about postcolonialism is equally true
about theories of globalization and about the entire tradition of critical
educational scholarship and activism. These reminders about insurgent
knowledges, however, need to be connected relationally to something
outside themselves.

Knowledge from Below

If one of the most powerful insights of the literature in critical pedagogy and in the growing turn toward theories of globalization and postcolonial perspectives is the valorization of knowledge from below, is
this sufficient? We know that the issue is not whether "the subaltern
speak," but whether they are listened to (Apple and Buras, 2006; Spivak,
1988). Yet this too can be largely a rhetorical claim unless it gets its
hands dirty with the material realities faced by all too many subaltern
peoples.

A focus within the critical community(ies) on "knowledge and voices
from below" has at times bordered on what Whitty called "romantic
possibilitarianism" (Whitty, 1974, 26–58). It is all so cultural that it
runs the risk of evacuating the gritty materialities of daily lives and of
economic relations. Yet with its brutally honest picture of what life is

like for millions, even billions, of people who live, attempt to exist, on the edge, Mike Davis' book, *Planet of Slums* (2006), demonstrates in no uncertain terms that without a serious recognition of ways in which the conjunctural *specifics* of the effects of global capital are transforming the landscape we sometimes too abstractly theorize about, we shall be unable to understand why people act in the ways they do in such situations. Work such as Davis's goes a long way toward correcting the over-emphasis on the discursive that so often plagues parts of postcolonial and critical pedagogical literature in education and elsewhere. And many of us need to be constantly reminded of the necessity to ground our work in a much more thorough understanding of the realities the oppressed face every day. Any work in education that is not grounded in these realities may turn out to be one more act of colonization.

Connecting with History

It is important to remember that in the Americas and elsewhere the positions inspired by, say, postcolonialism are actually not especially new in education. Even before the impressive and influential work of the great Brazilian educator, Paulo Freire (1970), subaltern groups had developed counter-hegemonic perspectives and an extensive set of ways of interrupting colonial dominance in education and in cultural struggles in general (see, e.g., Jules, 1991; Lewis, 1993, 2000; Wong, 2002; Livingston, 2003). But the fact that theories of globalization and post-colonialism are now becoming more popular in critical educational studies is partly due to the fact that the field itself in the United States and throughout Latin America, for example, has a very long tradition of engaging in analyses of hegemonic cultural form and content and in developing oppositional educational movements, policies, and practices (see, e.g., Apple, 2004; Apple, 2006a; Apple and Buras, 2006). But, as we know so well, the place that Freire has as both an activist in and theorist of these movements is unparalleled.

Thinking about Freire is more than a reminder of the past. It points to the continuing significance of Freire and Freirian-inspired work for large numbers of people throughout the world. While some have rightly or wrongly challenged the Freirian tradition and argued against a number of its tendencies (see, e.g., Au and Apple, 2007), the tradition out of which he came, which he developed throughout his life, and which continues to evolve is immensely resilient and powerful. While Freire's influence is ever-present, thinking about it has brought back some powerful recollections. Like others, I too had a history of interacting with

Paulo Freire. I hope that you will forgive me if I add a personal example of my own here, one that ratifies the respect so many people have for the man and his ideas.

Freire and Critical Education: A Personal Vignette

After delayed flights, I finally arrived in Sao Paulo. The word exhausted didn't come close to describing how I felt. But a shower and some rest weren't on the agenda. We hadn't seen each other for a while and Paulo was waiting for me to continue our ongoing discussions about what was happening in Sao Paulo, now that he was Secretary of Education there.

It may surprise some people to know that I was not influenced greatly by Paulo, at least not originally. I came out of a radical laborist and antiracist tradition in the United States that had developed its own critical pedagogic forms and methods of interruption of dominance. I had immense respect for him, however, even before I began going to Brazil in the mid-1980s to work with teachers unions and the Workers' Party. Perhaps it was the fact that my roots were in a different but still very similar set of radical traditions that made our public discussions so vibrant and compelling.

There were some areas where Freire and I disagreed. Indeed, I can remember the look of surprise on people's faces during one of our public dialogues when I supportively yet critically challenged some of his positions. And I can all too vividly remember the time when I had just gotten off those delayed flights and he and I quickly went to our scheduled joint seminar before a large group that had been waiting for us to arrive. The group was made up of the militants and progressive educators he had brought to work with him at the Ministry of Education offices in Sao Paulo. During the joint seminar, I worried out loud about some of the tactics that were being used to convince teachers to follow some of the Ministry policies. While I agreed with the Ministry's agenda and was a very strong supporter of Paulo's nearly herculean efforts, I said that—as a former president of a teachers union and as someone who had worked with teachers in Brazil for a number of years—there was a risk that the tactics being employed could backfire. He looked directly at me and said that he and I clearly disagreed about this.

The audience was silent and waiting—for distress, for "point scoring," for a break in our friendship? Instead what happened was one of the most detailed and intense discussions I have ever had in my life. For nearly three hours, we ranged over an entire terrain: theories about epistemologies; the realities of teachers' lives; the realities of life in favelas;

the politics of race and gender that needed to be dealt with seriously alongside class; the international and Brazilian economy; rightist media attacks on critical education in Brazil and on him personally; what strategies were needed to interrupt dominance in the society and in the daily lives of schools; his criticisms of my criticisms of their strategies; my suggestions for better tactics; the list could go on and on.

This wasn't a performance in masculinities, as so many public debates are. This was something that demonstrated to me once again why I respected him so much. There was no sense of "winning" or "losing" here. Paulo and I were fully engaged, wanting to think publicly, enjoying both the richness of our dialogue and our willingness (stimulated constantly by him) to enter into a field that required that we bring in *all* that we knew and believed. For him, and for me, education required the best of our intellectual and emotional resources. I'm not certain we ultimately resolved our disagreements. I know that I was taken with his passion and his willingness to listen carefully to my worries, worries based on my previous experiences with political/educational mobilization in other nations.

I also know that he took these issues very seriously (see, e.g., Apple, 1999). Perhaps a measure of this can be seen when, after that three-hour dialogue that seemed to go by in a flash, he had to leave for another meeting that had been delayed because of our discussion. As he and I said our goodbyes, he asked the audience to stay. He then asked me if I could stay for as long as it took so that the audience and I could continue the discussion at a more practical level. What could be done to deal with the concerns I had? Were there ways in which the people from the ministry and from the communities that were in the audience might lessen the risk of alienating teachers and some community members? What strategies might be used to create alliances over larger issues, even when there might be some disagreements over specific tactics and policies?

It is a measure of Paulo's ability, as a leader and as a model of how critical dialogue could go on, that another two hours went by with truly honest and serious discussion that led to creative solutions to a number of problems that were raised as people reflected on their experiences in favelas and in the ministry. This to me is the mark of a truly great teacher. Even when he wasn't there, his emphasis on honestly confronting the realities we faced, on carefully listening, and on using one's lived experiences to think critically about that reality and how it might be changed remained a powerful presence. He was able to powerfully theorize and to help others do the same because he was engaged in what can only be described as a form of praxis. I shall say more about the crucial importance of such concrete engagements in a later section of this chapter.

This was not the only time Paulo and I publicly interacted with each other. We had a number of such discussions in front of large audiences. Indeed, in preparation for writing this chapter, I took out the tape to listen to one of my public interactions with Freire. It reminded me that what I've said here cannot quite convey the personal presence and humility Paulo had. Nor can it convey how he brought out the best in me and others. One of the markers of greatness is how one deals with disagreement. And here, once again, Paulo demonstrated how special he was, thus giving us one more reason Paulo—friend, teacher, comrade—is still missed.

The Tasks of the Critical Scholar/Activist

But our task is not simply to be followers of Freire—or of any person for that matter. Here I am reminded of the radical sociologist Michael Burawoy's arguments for a critical sociology. As he says, a critical sociology is always grounded in two key questions: (1) "Sociology for whom?" and (2) "Sociology for what?" (Burawoy, 2005). The first asks us to think about repositioning ourselves so that we see the world through the eyes of the dispossessed. The second asks us to connect our work to the complex issues surrounding a society's moral compass, its means and ends.

For many people, their original impulses toward critical theoretical and political work in education were fueled by a passion for social justice, economic equality, human rights, sustainable environments, and an education that is worthy of its name—in short, a better world. Yet, this is increasingly difficult to maintain in the situation in which so many of us find ourselves. Ideologically and politically, much has changed. The early years of the twenty-first century have brought us unfettered capitalism that fuels market tyrannies and massive inequalities on a truly global scale (Davis, 2006). "Democracy" is resurgent at the same time, but it all too often becomes a thin veil for the interests of the globally and locally powerful and for disenfranchisement, mendacity, and national and international violence (Burawoy, 2005). The rhetoric of freedom and equality may have intensified, but there is unassailable evidence that there is ever deepening exploitation, domination, and inequality and that earlier gains in education, economic security, civil rights, and more are either being washed away or are under severe threat. The religion of the market (and it does function like a religion, since it does not seem to be amenable to empirical critiques) coupled with very different visions of what the state can and should do can be summarized in one word—neoliberalism (Burawoy, 2005), although we know that no single term actually can totally encompass the forms of dominance and subordination that have such long histories in so many regions of the world.

At the same time, in the social field of power called the academy—with its own hierarchies and disciplinary (and disciplining) techniques, the pursuit of academic credentials, bureaucratic and institutional rankings, tenure files, indeed the entire panoply of normalizing pressures surrounding institutions and careers—all of this seeks to ensure that we all think and act "correctly." Yet, the original impulse is never quite entirely vanquished (Burawoy, 2005). The spirit that animates critical work can never be totally subjected to rationalizing logics and processes. Try as the powerful might, it will not be extinguished—and it certainly remains alive in a good deal of the work in critical pedagogy.

Having said this—and having sincerely meant it—I need to be honest here as well. For me, some of the literature on "critical pedagogy" is a vexed one. Like the concept of postcolonialism, it too now suffers from a surfeit of meanings. It can mean anything from being responsive to one's students, on one hand, to powerfully reflexive forms of content and processes that radically challenge existing relations of exploitation and domination, on the other. And just like some of the literature on postcolonialism, the best parts of the writings on critical pedagogy are crucial challenges to our accepted ways of doing education.

But once again, there are portions of the literature in critical pedagogy that may also represent elements of conversion strategies by new middle-class actors who are seeking to carve out paths of mobility within the academy. The function of such (often disembodied) writing at times is to solve the personal crisis brought about by the "contradictory class location" (Wright, 1985) of academics who wish to portray themselves as politically engaged, but almost all of their political engagement is textual. Thus, their theories are (if you will forgive the use of a masculinist word) needlessly impenetrable, and the very difficult questions surrounding life in real institutions—and of what we should actually teach, how we should teach it, and how it should be evaluated—are seen as forms of "pollution," too pedestrian to deal with. This can degenerate into elitism, masquerading as radical theory. But serious theory about curriculum and pedagogy needs to be done in relation to its object. Indeed, this is not only a political imperative but an epistemological one as well. The development of critical theoretical resources is best done when it is dialectically and intimately connected to actual movements and struggles (Apple and Aasen, 2003; Apple, 2006a, 2006b).

Once again, what Michael Burawoy (2005) has called "organic public sociology" provides key elements of how we might think about ways of dealing with this here. In his words, but partly echoing Gramsci as well, the critical sociologist:

> ... works in close connection with a visible, thick, active, local, and often counter-public. [She or he works] with a labor movement, neighborhood association, communities of faith, immigrant rights groups, human rights organizations. Between the public sociologist and a public is a dialogue, a process of mutual education ... The project of such [organic] public sociologies is to make visible the invisible, to make the private public, to validate these organic connections as part of our sociological life. (Burawoy, 2005, 265)

This act of becoming (and this is a *project*, for one is *never* finished, *always* becoming) a critical scholar/activist is a complex one. Because of this, let me extend my earlier remarks about the role of critical research in education. My points here are tentative and certainly not exhaustive. But they are meant to begin a dialogue over just what it is that "we" should do.

In general, there are seven tasks in which critical analysis (and the critical analyst) in education must engage (Apple, 2006b).

1. It must "bear witness to negativity." That is, one of its primary functions is to illuminate the ways in which educational policy and practice are connected to the relations of exploitation and domination—and to struggles against such relations—in the larger society.
2. In engaging in such critical analyses, it also must point to contradictions and to spaces of possible action. Thus, its aim is to critically examine current realities with a conceptual/political framework that emphasizes the spaces in which counter-hegemonic actions can happen or are now going on.
3. At times, this also requires a redefinition of what counts as "research." Here I mean acting as "secretaries" to those groups of people and social movements who are now engaged in challenging existing relations of unequal power, or in what has been called elsewhere as "non-reformist reforms." This is exactly the task that

was taken on in the thick descriptions of critically democratic school practices in *Democratic Schools* (Apple and Beane, 2007) and in the critically supportive descriptions of the transformative reforms such as the Citizen School and participatory budgeting in Porto Alegre, Brazil (Apple and Aasen, 2003).

4. When Gramsci (1971) argued that one of the tasks of a truly counter-hegemonic education was not to throw out "elite knowledge" but to reconstruct its form and content so that it served genuinely progressive social needs, he provided a key to another role "organic intellectuals" might play (see also Apple, 1996; Gutstein, 2006). Thus, we should not be engaged in a process of what might be called "intellectual suicide." That is, there are serious intellectual (and pedagogic) skills in dealing with the histories and debates surrounding the epistemological, political, and educational issues involved in justifying what counts as important knowledge. These are not simple and inconsequential issues, and the practical and intellectual/political skills of dealing with them have been well-developed. However, they can atrophy if they are not used. We can give back these skills by employing them to assist communities in thinking about this, learn from them, and engage in the mutually pedagogic dialogues that enable decisions to be made in terms of both the short-term and long-term interests of oppressed peoples.

5. In the process, critical work has the task of keeping traditions of radical work alive. In the face of organized attacks on the "collective memories" of difference and struggle, attacks that make it increasingly difficult to retain academic and social legitimacy for multiple critical approaches that have proven so valuable in countering dominant narratives and relations, it is absolutely crucial that these traditions be kept alive, renewed, and when necessary criticized for their conceptual, empirical, historical, and political silences or limitations. This involves being cautious of reductionism and essentialism and asks us to pay attention to what Fraser (1997) has called both the politics of redistribution and the politics of recognition. This includes not only keeping theoretical, empirical, historical, and political traditions alive, but, very importantly, also extending and (supportively) criticizing them. And it also involves keeping alive the dreams, utopian visions, and "non-reformist reforms" that are so much a part of these radical traditions (Teitelbaum, 1993; Apple, 1995; Jacoby, 2005).

6. Keeping traditions alive and also supportively criticizing them when they are not adequate to deal with current realities cannot be done unless we ask "For whom are we keeping them alive?" and "How and in what form are they to be made available?" All of the things I have mentioned before in this tentative taxonomy of tasks require the relearning or development and use of varied or new skills of working at many levels with multiple groups. Thus, journalistic and media skills, academic and popular skills, and the ability to speak to very different audiences are increasingly crucial.

7. Finally, critical educators must *act* in concert with the progressive social movements their work supports, or in movements against the rightist assumptions and policies they critically analyze. Thus, scholarship in critical education or critical pedagogy does imply becoming an "organic intellectual" in the Gramscian sense of that term (Gramsci, 1971). One must participate in and give one's expertise to movements surrounding struggles over a politics of redistribution and a politics of recognition. It also implies learning from these social movements. This means that the role of the "unattached intelligentsia" (Mannheim, 1936), someone who "lives on the balcony" (Bakhtin, 1968), is not an appropriate model. As Bourdieu (2003) reminds us, for example, our intellectual efforts are crucial, but they "cannot stand aside, neutral and indifferent, from the struggles in which the future of the world is at stake" (11).

These seven tasks are demanding and no person can singly engage equally well in all of them simultaneously. What we can do is honestly continue our attempt to come to grips with the complex intellectual, personal, and political tensions and activities that respond to the demands of this role. Actually, although at times problematic, "identity" may be a more useful concept here. It is a better way to conceptualize the interplay among these tensions and positions, since it speaks to the possible multiple positionings one may have and the contradictory ideological forms that may be at work both within oneself and in any specific context (see Youdell, in process). And this requires a searching critical examination of one's own structural location, one's own overt and tacit political commitments, and one's own embodied actions once this recognition in all its complexities and contradictions is taken as seriously as it deserves.

This speaks to the larger issues about the politics of knowledge and people, of which I spoke earlier and to which postcolonial authors such as Young (2003), Bhabha (1994), Spivak (1988), and others have pointed. Concepts such as "hybridity," "marginalization," "subaltern," "cultural politics," and the entire panoply of postcolonial and critical pedagogic vocabulary can be used in multiple ways. They are meant to signify an intense set of complex and contradictory historical, geographic, economic, and cultural relations, experiences, and realities. But what must *not* be lost in the process of using them is the inherently *political* nature of their own history and interests. Used well, there is no "safe" or "neutral" way of mobilizing them—and rightly so. They are meant to be radically counter-hegemonic and also to challenge even how we think about and participate in counter-hegemonic movements. How can we understand this if we do not participate in such movements ourselves? Freire certainly did. So did E. P. Thompson, C. L. R. James, W. E. B. DuBois, Carter Woodson, and so many others. Can we do any less?

CHAPTER 2

Making Sense of the Call for Scientifically Based Research in Education

Ginette Delandshere

In an unprecedented move the U.S. government has essentially defined *scientifically based research* by law (The No Child Left Behind Act, 2001) as the use of *randomized controlled trials* to provide solid evidence of what works in education to policy makers, teachers, parents, researchers, and other consumers of research. This definition is now enacted and supported by a national and international review process (e.g., The Campbell Collaborative, The What Works Clearinghouse [WWC]) and by government funding practices (e.g., The Institute of Education Sciences [IES]) as well as other funding agencies. Scientifically based research, so defined, has been placed at center stage since the creation in 2000 of the Campbell Collaborative, an international "non-profit organization that aims to help people make well-informed decisions about the effects of interventions in the social, behavioral, and educational arenas" with the aim "to prepare, maintain and disseminate systematic reviews of studies of interventions" using randomized controlled trials (The Campbell Collaboration, 2000). The No Child Left Behind Act (2001) also makes more than a hundred references to scientifically based research and many websites of the state department of education contain lengthy documents to draw school districts' attention to the requirement of implementing scientifically based programs in the schools in order to receive federal funding. To follow suit, many government requests for research proposals in education or for evaluation of federally funded programs, in general, explicitly require the use of randomized controlled trials (or quasi-experimental designs, but only when randomization is not possible).

In 2002 a committee of the National Research Council (NRC) published a report on "Scientific Research in Education" (Shavelson and Towne, 2002). This report was an attempt to broaden the definition of scientifically based research that had been proposed by legislators to prescribe all those educational services and research that would now be funded with federal monies. Although the report focuses a great deal on the use of experimental designs, it does recognize the added contribution that such strategies as case study, ethnography, mixed method designs, and so on can make to education research. What prevails throughout the report is a strong emphasis on causal relationships, randomization, replication, measurement, and an exclusive post-positivist stance on research in education (Eisenhart and Towne, 2003) while explicitly rejecting postmodernism defined as "an extreme epistemological perspective that questions the rationality of the scientific enterprise altogether, and instead believes that all knowledge is based on sociological factors like power, influence and economic factors" (Shavelson and Towne, 2002, 25). Other than this exclusive stance, the authors of the report do not explicitly address epistemological issues, but the recurring opposition they make between scientific, objective, accurate versus the ideal, ideological, and value-based research is reminiscent of the positivist fact/value distinction and implicitly elevates this conception of scientific research over other forms of research or of knowing. The report appears to have had little impact on expanding the definition of scientifically based research, as it had intended, while the funding practices of the U.S. federal government requiring the use of randomized controlled trials in educational research have remained unaltered.

To reinforce this practice the U.S. government created the What Works Clearinghouse, an electronic library (a national analogue to the international Campbell Collaborative) containing reviews that identify "primarily well conducted randomized controlled trials and regression discontinuity studies, and secondarily quasi-experimental studies of especially strong design" (http://ies.ed.gov/ncee/wwc/overview/review. asp?ag=pi, accessed March 11, 2008) to provide solid evidence of what works in education.

The WWC was established in 2002, and launched in 2004, by the US Department of Education's IES to provide educators, policymakers, researchers, and the public with a central and trusted source of scientific evidence of what works in education. The WWC aims to promote informed education decision making through a set of easily accessible databases and user-friendly reports that provide education consumers with ongoing, high-quality reviews of the effectiveness of replicable educational interventions (programs, products, practices, and policies) that intend to improve student outcomes (http://ies.ed.gov/ncee/wwc/overview/, accessed March 11, 2008).

To be included in these reviews, studies are judged according to the following criteria reflecting an exclusive focus on the use of randomized controlled trials:

• Meets Evidence Standards—randomized controlled trials (RCTs) that do not have problems with randomization, attrition, or disruption, and regression discontinuity designs that do not have problems with attrition or disruption.
• Meets Evidence Standards with Reservations—strong quasi-experimental studies that have comparison groups and meet other WWC Evidence Standards, as well as randomized trials with randomization, attrition, or disruption problems and regression discontinuity designs with attrition or disruption problems.
• Does Not Meet Evidence Screens—studies that provide insufficient evidence of causal validity or are not relevant to the topic being reviewed. (http://ies.ed.gov/ncee/wwc/overview/review.asp?ag=pi, retrieved 3/11/2008)

Needless to say that reviews with such narrow criteria typically exclude more than 90 percent of the studies considered for each topic, and the final review reports only contain a few studies judged worthy of attention as a "trusted source of scientific evidence." To expand this systematic review of research the U.S. Department of Education also supported a number of other initiatives, such as the Best Evidence Encyclopedia (www.bestevidence.org) associated with the Center for Data-Driven Reform in Education at Johns Hopkins University and the Comprehensive School Reform Quality Center (www.csrq.org) at the American Institutes for Research (AIR).

This flurry of activity leaves no doubt as to the conception of scientifically based research that is now being promulgated by official sources. To understand the meaning of this prescriptive movement and the potential consequences that this may have for our work and for the project of education in general, I will analyze some of the assumptions that underlie the use of randomized controlled trials and the shortcomings of making this approach an exclusive condition for providing trusted sources of information on which to make policy decisions. I also consider some of the origins of this movement and question the wisdom of giving primacy to any research method or methodology.

Making Sense of the Scientifically Based Movement

The current definition of scientifically based research, and its sanction by official education agencies and to some extent by the academy, places important constraints on the work of researchers. Although there is nothing wrong with conducting experiments, placing this research methodology at center stage and presenting it as generating the most legitimate and exclusive forms of evidence as the basis for policy decisions has major marginalization effects that threaten educational research and the process of inquiry at its core. First, it marginalizes researchers who have different aims and use different approaches to inquiry as well as the representations and understandings that these yield. Second, in doing so it limits the kind of research questions that can be investigated, since not all research questions can be answered on the basis of randomized controlled trials, and this also limits the kind of evidence that can be brought to bear to further our understanding of the social world. Third, it positions participants as interchangeable objects of research randomly assigned to groups without consideration of their individual, social, and cultural experiences or allowing for their active participation in the inquiry process, a stance that I believe poses serious issues of representation as well as ethical concerns.

Administrative pressures on faculty members to compete for external funding has intensified at a time when public funding of universities has considerably decreased and other research monies have also become scarce (Lagemann, 1997, 2000). In this context funding practices that reward the use of randomized controlled trials, in effect, results in the marginalization of nonexperimental researchers who find themselves unsupported in their scholarly work. Even if they are able to continue their research unfunded, efforts to publish their work may not be as successful and are less likely to receive attention than the work of researchers who comply with the mainstream definition of scientifically based research. Marginalization affects researchers in different ways. It has direct effects on individual researchers through the reward system, but it also affects their participation in conversations and dialogues about the meaning and value of educational research. Elevating a particular definition of research to a privileged position makes questioning it difficult because other voices are silenced, a condition that directly threatens the very meaning of research and inquiry. If research can only be undertaken in a particular way, it also imposes a particular worldview that necessarily undermines other worldviews. This returns research to a highly convergent activity (Kuhn, 1970) conducted by a very homogeneous group of researchers, a condition that suppresses the possibility of new understandings and of varied representations and interpretations of the world (Delandshere, 2004). Since research serves to interpret the social realities it studies, and yields representations that in turn frame people's perception of their own reality, the current situation that seems to portray a particular conception of research as universal may result in a closed and unquestionable system of representations that is not conducive to a spirit of inquiry.

No one would argue with the need to support research claims with evidence, but the problem here is that the notion of scientific evidence called for has been reduced to that generated by a particular methodology, thereby excluding other types of evidence. The methodology itself privileges particular questions and effectively limits the possibility of answering other related and important questions. The question of what works to effectively improve student outcomes is narrow and yields partial representations of education, schooling, and learning. Questions that seek answers to what constitutes important student learning, what works for whom, how do students experience these tested interventions, or are these interventions worth implementing at all costs cannot be answered through the use of randomized controlled trials, and they would therefore not be regarded as legitimate research questions in studies financed by federal funds. Who will answer these important questions and who decides what the important questions are? With randomized controlled experiments, the types of evidence called for are those that reveal regularities, constancies, and systematic differences between groups, with little attention to the individuals who constitute these groups and the social and cultural experiences that they bring with them. Experiments deconstruct the complex into a few variables that are strictly manipulated, controlled, and that are presumably invested with the same meaning across all study participants. Complexity is regarded as an absence of control and random assignment as a way to deal with all the so-called nuisance variables that we do not know about, hence limiting the possibility of meaningful and useful explanations. Such studies may indeed provide causal descriptions based on systematic differences in group averages, but in and of themselves these differences do not provide causal explanations (Shadish, Cook, and Campbell, 2002).

Other inquiry strategies would be necessary to understand the why and the how of these differences. These studies also rest on a particular conception of the social as simply reflected in the aggregate of individual responses, and privilege psychological over social explanations of people's actions and behaviors (Danziger, 2000). Experimental studies leave many unanswered questions. Basing policy decisions exclusively on these studies would be problematic because they only consider the effects of a few manipulated variables in a strictly controlled environment, a condition that is very much unlike most contexts in which learning takes place. Science and causal inferences cannot, by themselves, dictate action (e.g., educational practices or policies) because actions based on research findings, as defined in the current context, do not address the morality of the actions taken, or the values and choices of groups and individuals (Delandshere, 2004).

Positioning research participants as interchangeable subjects of research randomly assigned to groups also yields particular and partial representations of their actions and behaviors. The beliefs and values held by participants, the cultural and social experiences they bring into the situation, their intentions, and the meaning they make of their circumstances are at best reduced to a few variables and most often ignored as important factors that can contribute to the explanations of their own actions. Instead, individuals' histories are assumed to average out through the random assignment process and therefore "everything else being equal" causal inferences can be made regarding the particular intervention being tested. This assumption has serious limitations given the partial representations and explanations that result from these studies. The ethical stance that underlies this positioning of research participants as subjects of research rests on a traditional code of ethics grounded in the principles of respect for individual autonomy and right of self-determination (an ethic of doing no harm that calls for informed consent of the participants, anonymity, privacy, and confidentiality) and the researchers' responsibilities to the research community. From a broader social ethics perspective, an ethical stance solely defined by respect for individual autonomy is problematic because it does not address the researchers' social responsibilities for and commitment to improving people's well-being. In addition, true informed consent is not really possible given the unpredictable nature of research, the process by which we obtain consent, the myth of anonymity and confidentiality, the problematic relations of power between researchers and participants, and the fact that some research practices may essentially be coercive (Eisner, 1991; Fine et al., 2000; Malone, 2003). For example, how can participants be truly informed of the purpose of conducting a particular study when researchers appear to have relinquished their responsibilities with regard to how research findings are used and the consequences these might have for the participants (a disposition that seems to characterize many researchers but particularly those engaged in experimental studies)?

Detached and objective observations of subjects characterize the work of researchers who use randomized controlled trials, a view that is consistent with the idea of neutral science. Science here is to be preoccupied with the idea of truth, not the idea of good; it is presumably value-free, or at least values are not addressed explicitly as a part of the studies. In this perspective there is a sense in which researchers are satisfied in constructing knowledge by conscientiously following the methodological protocols sanctioned by the scientifically based research movement, but they appear to have little concern for their social responsibilities and the social consequences of their work. This is particularly peculiar at a time when a number of scholars have advocated for researchers to take on much greater social responsibility and a deeper commitment to human well-being in their work. Recognizing that nonharmful research is not equivalent to good research, Hostetler (2005) states, "For their research to be deemed good in a strong sense, education researchers must be able to articulate some sound connection between their work and a robust and justifiable conception of human well-being" (16). Although he recognizes the importance of methodological debates, he believes that good research is ultimately an ethical issue and that questions of well-being should be foregrounded and "vigorously debated," a position that does not appear readily reconcilable with the methodological preoccupation with randomized controlled trials. Hostetler points to the complexity of the question of what is "good," but also to the importance of making it an object of inquiry and of developing "an ethical conception of what is good" (21). He concludes that as researchers we need to go beyond our traditional research questions and that

> [w]e need to think about how we can make life better for people. We need to think beyond our taken-for-granted ideas of well-being and what is good and make those ideas the objects of serious, communal inquiry. Serving people's well-being is a great challenge, but it is also our greatest calling. (21)

In treating research participants as interchangeable subjects and in ignoring the particulars of their life histories, an experimental methodology imposes implicit and taken-for-granted ideas of good on the research participants as well as particular representations of them in research writing. Although this is not the exclusive problem of experimentalist researchers, Fine (1994; 2000) has argued that much research suffers the same problem of speaking "of" and "for" others and of misrepresenting them; basing policy decisions and practice exclusively on evidence resulting from the use of their methodology will continue to impose particular conceptions of good without the opportunity to question or debate these.

Origins of Scientifically Based Research

Another way to make sense of the scientifically based research movement is to consider its history and the context in which it reemerged in the last decade. One could say that this scientific research movement came about because of politicians' frustration with the lack of consistent evidence on which to base policy decisions. Large-scale studies are expensive to conduct, and the scarcity of resources may partly explain why some researchers are turning to more theoretical rather than empirical work or are conducting smaller and more local studies. The quality of education research has also been questioned, and it is the case that many studies are less than adequate and do not often have explicitly articulated implications for policy or practice. Politicians' frustration may have led them to believe that mandating a particular research methodology would have solved the problem. This movement, however, did not originate from the NCLB legislation. Pressures to legislate the use of experiments for the evaluation of federally funded programs can be traced back to the 1970s when the U.S. Congress introduced a bill requesting a study of these program evaluations. The study was mandated by the 1978 Educational Amendments and led to several independent reports (Boruch and Cordray, 1982; Cronbach et al., 1980; Pincus, 1980; Raizen and Rossi, 1981). In a 1982 report to the U.S. Congress and the U.S. Department of Education, Boruch (who is the cochair of he Campbell Collaborative and was a member of the NRC committee that authored "Scientific Research in Education," as well as a principal investigator for the WWC) was already making explicit recommendations for the use of experiments in the conduct of evaluation and that these should be authorized by law (Boruch and Cordray, 1982 as cited in Boruch, 1982).

> We recommend that higher quality evaluation designs, especially randomized experiments, be authorized explicitly in law for testing

new programs, new variations on existing programs, and new program components. (1)

At this juncture it is important to realize how the neoliberal political and economic movement that was developing at the time may have framed the particular conception of scientific education research that is now being imposed on the research community. During the 1960s and 1970s a number of large federally funded education programs (e.g., the Elementary and Secondary Education Act, PL 94–142, Headstart) were implemented in the United States. Following this period, concerns about government expenditure and a perceived decline in the quality of education (e.g., A Nation At Risk, 1983) called into question the efficiency of some of these programs and resulted in mandated evaluation of all federally funded programs. Request for evidence, showing that government intervention was working to improve education, was mounting and eventually developed into the assessment and accountability movement that began in the 1980s and is still vigorously enforced today. Neoliberalism was founded on free-market economy as a means to achieve progress and economic growth, and presumably social justice, while rejecting government intervention. A parallel can be drawn here with the first liberalism (classical liberalism) era preceding WWII, which was guided by a blind faith in science and technology to engineer society and ensure progress and economic growth, a period characterized by an ideology of efficiency and certainty. Today the same faith in solid scientific evidence of what works to organize schooling, learning, and teaching (to ensure progress and economic growth) has forcefully reemerged and is reflected in the current scientifically based research mandate. Matters of education are more and more influenced by business roundtables that form partnerships with states and universities to shape the workforce for a global economy and meet market demands (Carnevale and Desrochers, 2003). Globalization has also created major uncertainties (e.g., clash of cultures, autonomy of nation-states, national border control, and local vs. global interests) and economic restructuring has resulted in major fiscal crises for many states, reducing public resources for social services including education (Burbules and Torres, 2000; Slaughter, 1988). Limited public resources, intense critique of public schools, and a push toward privatization are consistent with a free-market ideology that conceives of the purpose of education and schooling mainly as the education of the workforce. Educating a global workforce for a global economy also calls for "global knowledge," the so-called proven and solid causal claims presumably generalizable across contexts. In this global context there is a sense in which differences between regions, nations, and cultures need to be controlled because global market economy values homogeneity. Demand has to be predictable, and production and workforce have to be consistent, efficient, and flexible in all places regardless of local cultures, social values, or political contexts. From this perspective it is important to find that education interventions that are proven to work in reaching similar outcomes and a research methodology that ignores peoples' cultural and social affiliations, their values, beliefs, and interests are more likely to yield consistent findings and general claims. Defining scientific research in education in terms of randomized controlled trials is consistent with the global market ideology because it presumably allows for causal inferences and solid, generalizable, and efficient (that is, proven) solutions or interventions aimed at a particular goal, that is, a highly trained workforce (Delandshere, 2004).

This current search for certainty and generalizability is, however, quite disconcerting, because it imposes great homogeneity and particular values and interests on a world of differences without much concern for peoples' quality of life, and it positions scientific research exclusively in the service of global economic interests at a time when the idea of the common good is ignored or strictly equated with economic growth and technological progress. The metaphors of *producing, banking,* and *consuming* knowledge used by electronic libraries (e.g., the WWC, the Campbell Collaborative) illustrate the contradictions between the notion of common good and the economic production ideology that underlies its attainment, thus imposing a particular view of the common good. Important purposes of education, such as enhancing civic participation, promoting democratic ideals, and developing an appreciation for the good life, are excluded from the scientific research discourse because, by definition, these ideas raise questions of beliefs, interests, and values, which science presumably cannot incorporate. But when scientific research is placed in service of global economic interests, its control by the academy is forsaken by relinquishing the freedom of thought and autonomy necessary to research activity, which in essence threatens the very nature of research itself.

Future of the Scientifically Based Research Movement

At this point in time it is difficult to envision what will happen to this scientifically based research movement. Although the publication of the NRC report in 2002 raised major controversies, as reflected in presentations at national organization meetings (e.g., American Educational Research Association [AERA]) and in professional journal publications (e.g., Educational Researcher, November 2002 issue; Teachers College Record, January 2005 issue), discussions of scientifically based research have since considerably diminished. A quick search of the program for the 2008 AERA annual conference meeting returned no entry for this topic and just two references for randomized experiments. The WWC has also experienced a great deal of difficulty meeting its initial agenda for review of specific topics of interest in education. At the same time, some researchers are deploring "the prevalence of qualitative methodology at AERA's annual meeting" (Eckardt, 2007) and other observers (Hess and LoGerfo, 2006) are assailing educational research for its failure to address real problems in education and important policy questions. What seems to continue are funding practices by federal agencies, with other foundations following suit, that have placed a premium on the use of randomized controlled trials as a condition for research funding.

Shortly after the publication of the "Scientific Research in Education" report (Shavelson and Towe, 2002), the critique of this movement by the academy seems to have considerably diminished. A plausible hypothesis here is that as researchers we feel that once our critiques have been rendered we can then continue our own work, ignoring the mandate. Eckardt's (2007) observation of the prevalence of qualitative methodology at AERA might partially support this idea. Another hypothesis could be that we have simply accepted the mandate as inevitable and are complying with it in our scholarship and search for funding. A recent issue of the *Educational Researcher* (January/February 2008) was indeed devoted to "Perspectives on Evidence-Based Research in Education" not as a critique of the movement but rather on issues of bias, reliability, and differences in the review process and syntheses of studies of educational interventions conducted by different organizations (e.g., WWC, Best Evidence Encyclopedia). Most contributors to the issue discuss the factors that need to be taken into account to make this review process more meaningful, reliable, and less biased. With the possible exception of one or two contributors, this is a critique that originates from within and that presumes the value of such a review process or a compliance with the current definition of scientifically based research. In response to this movement, although without an explicit critique of it, other researchers have also rallied around the concept of mixed methods or mixed methodology (Johnson and Onwuegbuzie, 2004; Tashakkori and Teddlie, 2003) evidently as a means to accommodate and incorporate different methodologies and avoid their marginalization. Although this concept is not new and can be traced back to the 1980s, it has received renewed and vigorous attention in the recent years.

Whether through compliance with or critiques of the current mandates, or advocacy for combining existing methods, it seems that methodology has always been at the forefront of debates and divisions among education researchers. To be sure, methodological issues are important and should be debated, but placing a particular methodology, for that matter any methodology, at center stage with the exclusion of others appears terribly misguided. Rather, the important issues, problems, and questions as defined by stakeholders in particular contexts and conditions are what should be at center stage while methodologies as possible tools and strategies to inform how to proceed with inquiry should be situated at the margins, never to be imposed a priori on research questions, foci, and purposes. It is indeed important to remember that all methodologies frame inquiry questions or issues in particular ways, and hence they can only yield partial representations of the events or phenomena under study. All methodologies have limitations. The ongoing development and study of our methodological approaches are therefore critical to further our representations and understandings of the world around us. Such study is unlikely to occur, however, if one methodological perspective is deemed superior to all others.

CHAPTER 3

Limits to Knowledge with Existential Significance: An Outline for the Exploration of Post-secular Spirituality of Relevance to Qualitative Research

Phil Francis Carspecken

Introduction

This chapter is a truncated report on the second part of a three-part project I am currently working on. I am interested in post-secular frameworks for understanding spirituality, and its relevance to social research. By a post-secular understanding of spirituality, I mean an understanding of what is broadly referred to as spiritual in the West, but which emphasizes *limits to knowledge* and their existential significance. I do not mean anything having to do with metaphysics or religion or the idea of nonmaterial substances, beings, or entities aside from their importance as metaphors for what cannot be represented. I believe that this investigation has much relevance to social research, especially qualitative social research.

It can seem odd to refer to these interests with the term "post-secular," since the connection between spirituality and limits to knowledge can be found in texts as old as the Upanishads, some of which date between 900 and 600 B.C.E. They are insights that have been articulated in Western philosophical traditions that begin with Kant (1929) and continue in different forms through Fichte (1970), Hegel (1977),

Kierkegaard (1989), existentialism, critical theory, and poststructuralism. They are core features of Yogic philosophy, Chan and Zen Buddhism, and appear within some recent Christian theologies of intersubjectivity (see Theunissen, 2005).

But even though many insights on the relation of limits to knowledge and spirituality can be found in literature that spans so much time and so many cultures, the idea of a post-secular understanding of spirituality, which clearly references the secular features of modernity and implies that something can come next, still makes sense for a number of reasons. One is that even in the East, where spiritual concerns have been much more clearly articulated in terms of limits to knowledge than they have been in the West, spiritual issues are still by and large conveyed metaphorically via religious and metaphysical constructions. In the West this has been even more the case. Historically, with the rise of the "new learning" and natural philosophy during the 1600s and the Enlightenment that followed in the next century, Western culture managed to separate knowledge and institutions that specialize in various forms of "science"; dividing science from religion, morality, and aesthetics. This was a welcome movement for many reasons, even a form of freedom for human beings since knowledge became distinguished from the dictates of tradition and authority. By the nineteenth century, the West was moving toward a secular worldview at the level of culture relevant for overall social and system integration, with other things (including religion and spirituality) relocated as features of a diverse and tolerated private sector. One outcome of these historical changes was the rise of the ideology of scientism, which is the belief that all knowledge, to be knowledge rather than opinion or personalized beliefs, must take the form that it has in the spectacularly successful physical sciences. Because of the ideology of scientism, spirituality and many other aspects have been positioned as other-to-science, and consequently have been cut off from having a close relation to reason and rationality aside from that of a strict opposition. An understanding of spirituality in terms of limits to knowledge with existential significance is in this context appropriately named post-secular. It is continuous with reason and rationality, but becomes most noticeable at their boundaries.

In my ongoing large project, I am examining limits to knowledge in the physical sciences, limits to knowledge in the human-hermeneutic sciences, and the relation of both these sorts of limits to various spiritual teachings. I have already published a truncated report on the work from part one of this project (Carspecken, 2006). This chapter is an even more truncated report. What I am able to write in this chapter, given the amount of words allowed by the publishers, can do little more than introduce something of a basic idea of the second part of the larger project, and also point toward the third part. Many important arguments and issues cannot be addressed at all, and many things have to be simply asserted without elaboration, illustration, and proper argumentation.

Limits to Knowledge and Types of Knowledge

Although the principle of self-reflection risks the danger of solipsism, it is the very condition by which the world can turn into a world of objects. Self-reflection, then (and this is another of its major modern characteristics), makes mastery of the world dependent on the status of the world as a world of objects for a free and self-conscious subject who bears the promise of a free world. (Gasché, 1986, 14)

The expression *limits to knowledge* is here used to indicate enabling conditions for forms of knowledge that cannot themselves be known within the form they enable. Limits to knowledge do not mean that knowledge is finite, for there is no end to the knowledge that can be discovered or produced within a form that is entailed by limits such as these. But limits to knowledge specify the boundaries of a type of knowledge.

This idea has been well understood in Western philosophy since Kant (1929). But it is an understanding that can be achieved directly through reflective experience and the use of spiritual practices like meditation. In a book of his dialogues entitled *I Am That*, Nisargadatta (1973) answers the question "What am I?" as follows:

It is enough to know what you are not. You need not know what you are. For, as long as knowledge means description in terms of what is already known, perceptual, or conceptual, there can be no such thing as self-knowledge, for what you *are* cannot be described, except as total negation. All you can say is: "I am not this, I am not that." You cannot meaningfully say "this is what I am." It just makes no sense. What you can point out as "this" or "that" cannot be yourself. Surely, you cannot be "something" else. You are nothing perceivable, or imaginable. Yet, without you there can be neither perception nor imagination. You observe the heart feeling, the mind thinking, the body acting; the very act of perceiving shows that you are not what you perceive. (2)

One of the limits to knowledge we find in both the physical and hermeneutic sciences is the "I." It is necessary for a world of objects and events to differentiate within experience at all, and yet it cannot itself be an object of experience. Nisargadatta's (1973) response resembles Kant's (1929) discussion of the I in his *Critique of Pure Reason,* "in the synthetic original unity of apperception I am conscious of myself not as I appear to myself, nor as I am in myself, but only that I am"[1] (157).

When we think of knowledge in terms of objects and events that can be observed and modeled, which is basically the form in which our physical sciences operate, a number of conditions that enable this framework are missing from what can be understood within its boundaries. The I is one of them. It takes a reflection to discover these necessary conditions, and even then the I that is discovered is not discovered as something positive, not as an object or an appearance or an entity of any sort but rather a condition for the possibility of any object or appearance at all. So the I, and reflection itself, enters into representational frameworks as the negative. Dialectical philosophies like Hegel's (1977) are based on this insight, which is also central to teachings we find in Buddhism, Hinduism, and yogic teachings like Nisargadatta's (1973).

Scientism is possible through ignorance of the enabling conditions for the physical sciences. In my full project, and partly in the 2006 publication mentioned above, there are explorations of some of the possible linkages between critical philosophy and what Penrose (1994) calls the **X**-mysteries, "paradox mysteries" (237), of quantum physics. Is it possible, for example, that Nisargadatta's teaching, and Kant's, can enlighten us with respect to epistemological issues in physics? Particle physicist and Nobel Laureate Steven Weinberg (2001) explains one of these issues as follows:

> Much as we would like to take a unified view of nature, we keep encountering a stubborn duality in the role of intelligent life in the universe, as both the observer of nature and part of what is observed. (77)
>
> At present, we do not understand even in principle how to calculate or interpret the wave function of the universe, and we cannot resolve these problems by requiring that all experiments give sensible results, because by definition there is no observer outside the universe who can experiment on it. (79)

Although much is illuminated about limits to knowledge when *reflection* is appealed to, this approach is itself limited if reflection is understood within the paradigm of the subject-object relationship. Scientism takes knowledge in the form of a subject-object relationship and then forgets about the subject side, trying to turn back to it with physical models of consciousness that are in the end self-contradictory. Dialectics and deconstruction both cast most of their arguments from within a bottom-line subject-object paradigm to give us crucial insights that are yet limited in their expression. But the subject-object paradigm itself comes from within a subject-subject paradigm: *intersubjectivity*.

In this regard, a stunning breakthrough, to my mind, occurred with the publication of *Knowledge and Human Interests* (*KHI*) by Jürgen Habermas (1968). Limits to knowledge in the physical sciences, and in those social sciences that take an objectivating perspective on human beings and social phenomena, which together are what he then called "empirical-analytic sciences," turn out to presuppose another domain of knowledge: what Habermas (1968) called "historical-hermeneutic" knowledge. Historical-hermeneutic knowledge, in turn, presupposes the possibility of a third type of knowledge and an active human interest related to it: critical knowledge corresponding to an interest in emancipation.

There are problems with the approach Habermas (1981, 1987) developed in *KHI*, which he overcame in his later work, *The Theory of Communicative Action* (TCA). In TCA we begin with communicative rationality and intersubjectivity. Within the structures of communicative rationality specialized forms of knowledge can be, and have been, developed, such as knowledge of our physical world and hermeneutic knowledge. I find TCA a better framework for investigating knowledge and its limits, but many of the striking insights one finds in *KHI* are obscured within Habermas's later work. So let's first consider Habermas's division of knowledge into types whose linkage to each other can be discovered through reflection. I only have space to present a condensed summary in the table below (table 3.1), and I have added in the table a number of things that Habermas may well not accept—particularly with respect to his third type of knowledge, the emancipatory-critical type.

Scientism is the ideology that all knowledge, to be knowledge and not opinion or mere belief, has to possess only the characteristics shown in the first column of the above table (Empirical-Analytic). Within scientism, the various factors set out in the other two columns of the table are believed to be explicable within the framework of the first column. Theories of the self and consciousness often reduce these concepts to physical forces, entities, and/or systems; communication is modeled in terms of physical processes or physical-like processes; human motivation is reduced to desires experienced by individual actors for certain objective or internally felt states; moral theories are based on how a collection of desires and needs, essentially different between individuals, can be

Table 3.1 Characteristics of types of knowledge in *KHI*

KHI *Type of Knowledge*	*Empirical-Analytic*	*Historical-Hermeneutic*	*Critical-Emancipatory*
Motivations	For having or avoiding tangible, objective states of affairs. For explaining what "is" in a world of objects and events.	For reaching agreements to coordinate actions. For understanding why others do as they do, what they experience, how they think. For being understood to meet basic recognition needs.	For understanding one's self. For becoming more of what one can become. For freedom and autonomy.
Actions	Instrumental	Communicative	Reflection for its own sake, and/or desire for desire, and/or nonaction.
Forms of knowledge and language	Copy knowledge, Models. Formalized languages with all rules and meaning of terms explicit. Meaning of terms reduced to measurement procedures.	Hermeneutically achieved explications of what was formerly implicit knowledge. Ordinary language always dependent on implicitly understood and changing rules of use and meanings.	Internal recognition. Necessarily noncommunicable states and experiences. Language use necessarily contradicts knowledge.
Relation of knowledge to being	Knowledge and being are separated and distinct. Knowledge is of reality. New knowledge does not change reality. Relations in reality are causal/correlational.	Knowledge is of knowledge and is self-transcending. New knowledge changes reality. Relations between people are illocutionarily maintained but usually in distortion via power.	Knowledge is fused with being. Being as knowledge of itself. It makes no sense to speak of relations within reality.
Constitution of the "Knower"	The self as anonymous universal observer and instrumental actor. Self transcends object domain as Kant's I.	Individuated self with unique autobiography. Self transcends itself as author and critic of one's self-narratives.	Self is pure reflection. Knower is its process of knowing itself. Self is the loved Other of God, and/or there is no self.
Relation to reflection	Reflection is restricted such that the form of inquiry cannot study itself in its own framework.	Reflection is internal to hermeneutics such that this form of inquiry can consider itself.	Reflection is the method and object of study simultaneously.
Validity	Based on the degree to which predictions are successful.	Based on the degree to which insiders recognize explicit articulations as something already known in some way.	Continuation of a self-formative process, and/or noncommunicable state of enlightenment.

Note: *KHI* = Knowledge and Human Interests

coordinated; and so on. Self-contradictions arise when these reductions are attempted, but there is no space to illustrate that concern here.

Conditions and Characteristics of the Hermeneutic Domain

One of Habermas's great insights in *KHI* is to reveal the location of those conditions that enable empirical-analytic science, but which cannot be themselves understood through empirical-analytic science, to be within the hermeneutic domain of knowledge rather than within a purely transcendental realm. Column one (Empirical-Analytic) in table 3.1 depends on and presupposes the other two columns. Within a theory of knowledge that takes intersubjectivity as primary (the subject-subject relationship, rather than the subject-object relationship), we can see how empirical-analytic science becomes enabled through a set of restrictions made within full human experience and knowing—the domain in which hermeneutic inquiry takes place. I cannot show all the interesting details here, but simply stated: when we remove the restrictions enabling empirical-analytic knowledge, we find ourselves to be and to have always already been within the domain of hermeneutic inquiry.

By hermeneutic inquiry I do not mean one form of qualitative research, as if phenomenological, constructivist, poststructuralist, critical, and other so-called paradigms or forms of qualitative research did not fundamentally employ hermeneutics. Hermeneutics is a theory that purports to tell us what we already implicitly know—how to understand meaning. It is an articulation of the way meaning is understood in ordinary everyday life, such that once articulated, the process can be made methodical and used more consciously in order to attain better understandings of meaning and avoid misunderstandings. All qualitative social researchers seek to understand meanings. The form of knowledge we produce is not, in the first instance, that of models but of reconstructions. Validity is not based on making accurate predictions but on the degree to which participants recognize our formulations. I cannot produce a full argument for this here, so I will simply state my contention that all qualitative researchers work within the conditions and characteristics presented in column two (Historical-Hermeneutic) in the above table, and many qualitative researchers concern themselves with features of column three (Critical-Emancipatory) as well. Moreover, all features of the hermeneutic process of understanding meaning can be understood through communicative action theory and the structures of intersubjectivity, including the famous hermeneutic circle.

Knowledge in the hermeneutic domain has both epistemological and ontological significance. This is one of the key things that distinguish hermeneutic knowledge from empirical-analytic data. Here, I can only concentrate on two of the many features of this connection between knowing and being when knowing is unrestricted: the relation between knowledge and human motivation, and the relation between knowledge and the knower.

Knowledge and desire. In row one (Motivation) of table 3.1 we see a progression in the relation of human motivation to knowledge: from that of a strong separation between goals, knowledge, and satisfaction of goals (so that knowledge takes the instrumental form of specifying means for tangible-objective ends); through that of a range of communicatively structured goals, some of which involve understanding and being understood for the sake of maintaining social identities and ontological securities; to that of a fusion between means and ends (in which knowledge takes the form of a reflection that simultaneously advances self-formative processes).

Knowledge and reality. We move in row four (Relation of knowledge to being) from a constituted form of reality, being, which is independent of the knowledge that may be had *of* it; through a form of being in which knowledge is both a social and personal feature of reality itself and knowledge of these knowledges; to a form of knowledge whose knowledge of itself is a form of being.

Knowledge and the knower. We move in row five (Constitution of the "Knower") from a knower or self that is a universal perceiver and instrumental actor; through a form of knower that is particularized, individuated via her capacity to tell stories of herself and also *not* be the character in those stories through being their author; to a form of knower that is its own process of knowing itself.

Limitations to Knowledge in the Hermeneutic Domain

A key set of *conditions* for hermeneutic knowledge pertains to what we find in column two (Historical-Hermeneutic) with respect to the motivations related to knowledge: the constitution of being and the constitution of the knower. A key set of limitations associated with these conditions gives us the associated characteristics listed in column three (Critical-Emancipatory) for each of these domains: desire does not occur within a subject that differs from the object of desire but is directed toward itself; being is knowledge of itself; the knower is its own process of knowing itself. Most of these "characteristics" involve reflexivity to such a high degree that, as formulated, they can seem self-contradictory or nonsense. All of them have been central to various spiritual teachings spanning recorded human history across many diverse human cultures. Efforts have been made to deal with them philosophically with dialectics (Fichte, 1970; Hegel, 1977; and others). In some forms of poststructuralism, particularly Derrida's, these characteristics have been deconstructed so as to leave the core concepts of "being," "self," "experience," and so on undermined but at the same time irreplaceable.

The contradictory and/or exceedingly dialectical features of these characteristics of knowledge that mostly pertain to human desires for freedom, love, emancipation, self-validation, certainty, self-actualization, realization, and so on occur because each has to do with limits to knowledge precisely where knowledge is a feature of ontology. They would not be of more than academic interest to us if they did not possess existential significance. But they do possess existential significance and are aspects of the human condition that all people live with, and, thus, of all people qualitative researchers work with. An effort to understand these features as limits to knowledge in the hermeneutic domain raises many aspects that concern qualitative researchers. It helps them in defending these very domains from the oppressions of scientism and scientistically ordered forms of life. It helps in understanding both what Habermas (1984, 1987) originally called the emancipatory interest and all the various forces and conditions that humans desire emancipation from. It also helps in producing ever more subtle understandings of human identity and its relation to human needs and desires.

The Existential Significance of Limits to Knowledge in the Hermeneutic Domain

I have room to make only a few points in this last section of the chapter. These will have to do with the character of human being as a desired, communicatively structured, and yet communicatively non-satisfiable process of self-knowing. The conditions that make a human self possible are the structures of intersubjectivity associated with the subject-subject relation. And these entail limits to knowledge that are existential in nature. Such conditions both pertain to a domain of human need that has been addressed in spiritual traditions and structure the relations between culture and identity upon which cultural power operates. Existentially, there is something essentially nonrepresentable and unknowable about all of us, which is paradoxically tied to self-dependencies on representations. Cultural power works by restricting the milieu through which we humans could otherwise attain at least our partial forms of self-knowledge freely so as to, among other things, discover these existential limits. Spirituality is therefore also often a feature of many discourses that resist forms of oppression and domination.

Freedom: The Subject as a Chronic Claim we Make and Simultaneously Transcend

In a book called *Losing the Moon*, Byron Katie (1998) tells her interviewers:

> People love to pretend they love the moon. They love to pretend there is one. They must—they're doing it. Look at the moon! And another truth is no one has ever looked up at the moon and believed it! We're just trying to get other people to continue to validate that it's up there at all. This is about the only reason we ask. It's just like a joining. "Isn't the moon beautiful! That's a true concept isn't it? Please validate that." (97)

We can add to Byron Katie's comments, "Please validate *me*!," for reasons I am about to explain. Meaning is a matter of claims made implicitly and explicitly by a human subject in relation to various audiences (other subjects/other subject positions). A great insight made by Gottlieb Frege was that truth is internally related to meaning; that when we understand the meaning of an assertion we understand (implicitly or explicitly) the conditions that would make the assertion true or false (Dummett, 1991). In his theory of communicative action, Habermas (1984, 1987) expands this insight from sentences that make assertions, based on Frege's discovery, to full speech acts that simultaneously carry claims divisible into the subjective, normative, and objective categories (expanding Frege via speech-act theory). When we understand an act of meaning we at least implicitly understand a cluster of claims made by the actor and the conditions in which these claims would be valid or invalid. This can be used to guide meaning reconstruction in qualitative research through what I have called the "validity horizon" (Carspecken, 1996, 2001, 2003). Knowledge is always an uncertain matter of claims, and a shared position with respect to a deep structure of culturally common claims is one of the factors that coordinates social action. Validity horizons also contain identity claims. A meaningful act also claims the existence and identity of the actor, and hence reconstruction of meaning in a validity horizon will include articulation of possible identity claims (Carspecken, 2003).

In his essay on universal pragmatics, Habermas (1998) notes that by acting meaningfully an actor demarcates herself from the three formally constituted worlds of objectivity, normativity, and subjectivity. The subject herself is not in any of these formal worlds but is demarcated from them through her acts. She, her "I," is demarcated such that we humans interact with each other with the necessary assumption that each of us is considered the intentional source of our actions and can and should be held responsible for what we do. Wilhelm Dilthey (2002) noted that the I, which Kant associated with the unity of apperception in sense experience, is replaced by the autobiography when we examine the unity of fully lived experience, not just sense. Hence, to be a person is to be capable of providing at least partial self-stories (an integration of our claims to be a "me"). But what we are is not exhausted by our own character in these stories—it is both these characters and we as their authors (a distinction between the I and the me). We transcend our identity claims as soon as we make them. We transcend our self-narratives as soon as we narrate them (Habermas, 1968). Human identity is a matter of "I-me" relations that we can always integrate into self-narratives, which, however, never fully capture what is represented of the self. I-me relations vary greatly in form from person to person and culture to culture, as do narrative genres, but because of intersubjectivity itself we always find the distinctions between I, me, and autobiography (see Huang, 2008; Winkle-Wagner, 2006 for innovative reconstructions of I-me relations in identity claims and self-narratives).

Freedom is an essential condition of being a person directly related to a dialectic of self-positing (claiming) in every meaningful act, self-knowing that follows self-positing, and the transcendence of the I that accompanies both.

Illocutionary Force, the Other and Human Needs Related to Self-knowing

Freedom is also a condition for illocutionary force. Patterns in human social life are attributable to many things, including power relations and the structure of objectified social environments like markets, which coordinate large numbers of purely instrumental actions. What distinguishes the human social domain from the natural domain of objects and events, however, is the coordination of action through illocutionarily attained agreements on criticizable validity claims. To grasp this fully, and its far-ranging significance, is to begin to grasp what is at the heart of Habermas' (1984, 1987) theory of communicative action. The validity claims internal to meaning can always, in principle, be consented to for no other reason than the ones their author would supply in their support. This is illocutionary force, a rational force that is structured in part by the freedom of actors to say yes or no to the claims made by another (see Habermas, 1998, for a full discussion of illocutionary force).

Now, for our interests, it becomes illuminating to ask the question of what would compel someone to agree to the specifically moral and normative claims of another actor (we should do such and such, not do such and such) when the consequences of an agreement conflict with various desires and fears. Why honor an agreement, for example, when doing so results in financial loss, or humiliation, or perhaps even physical injury? Normative and moral claims are essentially rooted in the need for positive regard from possible others, which, via developmental processes, comes to include positive regard from ourselves. There is no me to the self without norms, and no norms without generalized other positions that a person internally takes in relation to herself. As Robert Brandom (1994) basically points out from within his distinctive line of argumentation, at the end of the day the grounding of norms resides in a human need to find herself trustworthy from at least some of all possible other subject positions she has internalized. This becomes particularly clear when we consider moral actions made, even at the cost of consequences conflicting with other desires and fears, particularly when one is all alone. Why do not we steal something we want or need when no one would ever know about it? The reason is existential—the me part of the self only exists through our own position-taking with possible other subject positions, and we find ourselves through this intersubjective reflection as our self-knowing. The more we act so that we can approve of ourselves, the more subject-like we are. The more we act against this to favor other desires and fears, the more object-like we are. The limit-case of being object-like is that of losing the self. As Kierkegaard (1989) writes in a different but related context:

> The biggest danger, that of losing oneself, can pass off in the world as quietly as if it were nothing; every other loss, an arm, a leg, five dollars, a wife, etc. is bound to be noticed. (62–63)

Possible other subject positions are *culturally contingent* generalized other positions, of course, but due to the chronic transcendence of the I, the dialectics of self-knowing through possible other subject positions, and other factors as well, there is an internal telos to any generalized other that refines it toward the principle of any sentient Other in general.

I cannot fully explain the reasons in this short chapter, but I believe this is related to monotheistic notions of God. When those who believe in a monotheistic notion of God pray, they move toward the limit-case of communication represented by an absolute other subject who is also considered omniscient. Omniscience means that nothing need be said or thought to make one's self understood to the Other considered omniscient, and in fact saying anything at all can only distort what is meant. Hence prayer, as St. Augustine came close to noting in his *Confessions*, has the logic of leading toward full openness to the Other in silence (something I explore more in the larger project). Augustine (1991) wrote: "Indeed, Lord, to your eyes the very depths of a man's conscience are exposed, and there is nothing in me that I could keep secret from you, even if I did not want to confess it. I should not be hiding myself from you, but you from myself."

Contradictions and Limit Cases

The desire to know ourselves is fundamental to existing as a person, but because of the I-me structure of human identity it can never be fulfilled communicatively. What is desired is the recognition of other subjects *who are free* to affirm or negate us outside our control. Even if other subjects do affirm our identity claims, we can never be sure they really affirm us because we do not have access to their subjective states, nor can their apparent affirmations fully affirm us because they have limited access to our states and do not fully know what they affirm. This also means we cannot know and recognize ourselves, because we only have a perspective on ourselves that consists of internalized, generalized possible other subject positions, none of which is actually fully within our reach and all of which only provide contingent responses to the finite me part of our claims. The desire to *be* involves a form of knowledge that cannot satisfy this desire. Spiritual practices involve the limit conditions of this structure: a God who is the one and only other subject that can know us is one such limit condition, but it is one that, as St. Augustine (1991) seemed to know, would end communicative action in silent contemplation of an ultimate and omniscient Other. Meditative and deconstructive practices work in the other direction of limit cases: there is no self, at the end of the day, to be recognized.

The chronic nature of the identity claim, such that it is implicitly a feature of every meaningful act, can be discovered in such a way that one wishes to be free of it, to end their identity claims. I have no room to explain the idea, but there is always something false in a me-claim and, in addition, me-claims have implications for the identities of other people because they draw upon cultural identity-sets that use contrasts and comparisons (to claim to be one kind of a person means to not be another kind of person, which has implications for other actual persons). We can experience a desire to stop the process but will find that we are stuck. "It is impossible to get rid of one's self. But one wants to get rid of one's self both in despairingly willing to be oneself and in not willing to be oneself" (Theunissen, 2005, 14).

Knowledge and the Morality of Representing Other Subjects

Wilhelm Dilthey was very insightful to note that meaningful action has to both promote understandings between people and yet keep people existentially unique and different from each other (see Habermas, 1968, 166 for a brilliant discussion of this). No acts of meaning, then, can be completely understood by others. If others think they completely understand us, it is oppressive because our subject-status is then lost under the reifying gaze of someone else. Whole discourse-practices are capable of subjugating groups of people by claiming to know them totally in objectifying ways (hence Foucault's studies of power-knowledge fusions). So do ideologies, such as patriarchy and racism and homophobia. When those who are subjugated internalize the objectivating position of the Other for their own identity claims and self-reflections, then they are oppressed at a deeply existential and spiritual level.

Katie (1998) insightfully connects our experience of love with acknowledging dependence on others and respect for what we cannot ever know of others: "The 'me' you're speaking of does not exist—I remain only your story. I don't believe anything about you—that's love. Without a story—love" (54).

Qualitative researchers can bring this understanding of existentially crucial limits of knowledge to their research practices, following something like the poststructuralist "principle of antirepresentationalism" (May, 2004, 48). When representing other people, care must be taken to avoid hierarchies in the representational systems. Representing other human beings must allow these others to speak for themselves and must include the dignity and respect we give to others by acknowledging that we cannot fully understand or represent them.

Conclusions

Writing this chapter was frustrating for me because I found it so difficult to cope with the limited number of words allotted to it. I have tried to convey something of the entire project I am involved with. We can understand limits to knowledge as conditions that enable forms of knowledge that cannot themselves be understood within the form of knowledge enabled. Limits to knowledge in the physical sciences and in those human sciences that take an objectifying view of human beings and societies can be shown to be the results of restrictions made within a larger intersubjective/hermeneutic domain. This sort of analysis, pioneered by Habermas (1968) in *Knowledge and Human Interests*, clarifies differences between empirical-analytic forms of inquiry and hermeneutic forms of inquiry. To me, it seems that this alone is important and not well understood in our qualitative research communities.

But, in addition, knowledge and existence begin to come together within the hermeneutic domain, and they fuse at the limits of this domain where communicative practices must stop, or be used entirely in indirect ways. This is where we find post-secular forms of spirituality as a matter of limits to knowledge with existential significance. I will have more to say in my completed project about how an understanding of limits to knowledge can be integrated with research practices and even our very conception of research itself. At bottom it all suggests, strongly for me, the advantages of starting to talk about spirituality from a post-secular orientation: spirituality as it is entailed with those limits to human knowledge that constitute our existence as subjects rather than objects.

Note

1. Kant goes on to say that there is an experience of the "I" within inner-sense, but this does not give us what the "I" is "in-itself." The experience of the "I" within inner-sense seems to me to be what is called the "I-feeling" in Eastern thought—a "feeling" that is already gone as soon as it is noticed. After Kant's discussion, Hegel, Kierkegaard, and others discussed the "I" as the *negative* and as reflection itself, never as an object or entity but necessary for there to be any objects or entities.

CHAPTER 4

Theory of the Margins: Liberating Research in Education

Barbara Dennis

The Practice and Theory of Emancipatory Research

Galileo studied physics through observation and experimentation rather than through moral and verbal instruction, as was the Western European tradition in the sixteenth century. Arguing against the politically powerful Catholic perspective of the time, Galileo supported Copernicus' theory that the Earth revolved around the Sun. As he was opposing the mainstream knowledge of the time, Galileo was marginalized. Throughout Western history there have been people whose research was marginalized, those who did research in the fray. In some cases their work moved to the center (as with Galileo), in other cases it did not. Sometimes research is marginalized because its findings challenge the status quo in ways that are just not accepted. At other times, research is marginalized because its methods are not considered valid or reliable. In yet other times, research is marginalized because it attends to questions and persons whose life experiences are not considered important or worthy of research by those invested in the mainstream. Galileo's research not only offered new facts from which to contemplate the workings of the universe, but also liberated our scientific investigations from the hegemony of existing European religious dogma. The focus of this chapter involves describing and promoting a link between research that is conducted at the margins and the potential for emancipatory effects. Researchers concerned with liberatory/emancipatory goals hope that their work contributes to a freer, more democratic, and more egalitarian society. In this chapter, I lay out an argument suggesting that research *should* be (not just *can* be) both liberatory and factual and that this integrated goal is central to research at the margins (see Hammersley, 2005, for a counter argument). In other words, to meet research goals that integrate factual and emancipatory interests, it is necessary to include nonmainstream research.

To make this argument, I begin by describing the link between research at the margins and liberatory intentions. I describe this link on three levels—factual, meta-theoretical, and moral. The bulk of the chapter is devoted to justifying the need for nonmainstream research for purposes of emancipating social actors from lives and conditions that perpetuate inequalities and inhumanities. Both practical and theoretical justifications are provided. In the end, it is hoped that readers will both appreciate the liberatory potential of research and understand the extent to which research at the margins is necessary for manifesting that potential.

Marginalized and Liberatory Research: The Link

Doing research that is marginalized reflects a relation to the mainstream or center of a particular research community. That is, to talk about the margins at all is to talk about the contingent and political nature of research that contributes to centralizing some research over others regardless of whether one is engaged with quantitative or qualitative methods. In contrast to Weber's (1949) claim that social science involved a "disinterested" pursuit of truth, the identification of the margins of research activity involves specifically locating interest positions of inquiry (both those at the edge and those at the center). The center is not a neutral space (Korth, 2005). So long as the orthodoxy of research serves the interests of a particular social class, we have the confluence of ideology and truth at work in the research itself (Marx, 1970). Research at the margins, by its very nature, steps outside the mainstream in some way and thereby possesses, at least minimally, a form of liberation from the ideological force of the center.

One might engage in research at the margins by actively challenging the socially situated research orthodoxy, but it is also possible that one might simply find that one's scholarship is outside that orthodoxy. Doing research that involves critical, emancipatory aims also involves challenging the inequalities and ideological aspects of the status quo and criticizing aspects of research that, for whatever reason, retain and promote inequity, injustice, or disparity amongst humans in a given time and place. Just by working at the margins, a critique of research hegemony is lodged (see, e.g., Gilligan (1977) who studied women instead of men), as has been evidenced across a variety of methodological decisions inclusive of quantitative, qualitative, historical, theoretical, and mixed methodological approaches.

In this section of the chapter, I want to discuss how it is that marginalized research is linked to emancipatory goals. I will do that by describing the factual, meta-theoretical, and moral aspects of the link between doing research at the margins and the potential for social liberation through research.

Factual Relations

Research at the margins can help correct factual errors produced within the mainstream and thereby liberate practice and knowledge from inaccuracies. For example, coronary heart disease (CHD) accounts for more deaths of American women than any other single cause (and kills more women than men in the United States), but over the past twenty years, research on the disease has primarily excluded women (Grady et al., 2003a, 2003b). Women are a marginalized population for those involved with this particular research endeavor. The few studies that have included women have been reporting important CHD differences related to gender. This basically means that the findings of studies conducted with male subjects cannot be generalized to the medical treatment of women, as in fact they have been (e.g., Agency for Healthcare Research and Quality, 2006; Pope et al., 2000). It took studies at the gendered margins of this particular research community to correct the so-called medical facts and contribute to the emancipation of treatment and knowledge heretofore limited by male-centered research practices (Agency for Healthcare Research and Quality, 2006).

This is not a suggestion that research at the margins is more likely to be factually correct than research at the center. Nevertheless, most research at the margins is considered such because there is already mainstream research that is central to the research community at hand. Thus, the facts produced are already set in the context of mainstream facts. Discovering factual errors related to unexpected bias in the mainstream is an important manifestation of the liberatory potential for research at the margins.

Meta-theoretical Relations

Critical theories of various sorts can be used to establish a meta-theoretical link between doing research at the margins and engaging in emancipatory research purposes. Critical theories (such as those proposed through the works of Foucault, Derrida, and Habermas) are, in the first place, a critique of existing social science methodologies. In each case, the critique undermines a sense of power, unity, or truth by revealing the necessary contingencies found already in place. According to Habermas (1984, 1987), when a researcher claims certain things to be true, she does so with an assumption that others will be able to recognize the validity of the claims or understand on what grounds it is possible to critique or challenge the claims. One important way to find the contingent nature of a claim to truth is to see if it holds at the margins. Metatheoretically speaking, emancipation involves the critique of taken-for-granted assumptions of power, unity, and truth, with the margins serving as a source of critique and alternative perspective. The truth dialogue, so to speak, must be inclusive of counter-experiences and stories, meaning stories that do not fit and experiences on the edge of the status quo.

The differences between the margins and the mainstream supply the *starting content* for a critique. The structure of the relations between the margins and the mainstream also provide a *pathway* for critique. Benhabib (1992), Minh-ha (1989), and others (Habermas, 1998 and Freire, 1977, for examples) have argued that such critiques are necessary for emancipation from inequity, oppression, dogma, distorted ideology, and other modes through which people are dehumanized and limited in their potential for becoming fully human. Critical metatheory suggests that research is both a factual and moral endeavor whose work in society must include those at the margins.

When people offer reasons for their actions and seek understanding for their lives in dialogue with someone who differs from them, the demand to articulate taken-for-granted assumptions is present in principle. Studies on "whiteness" in the United States provide an interesting example. White folks in the United States began to have their race and ethnicity questioned once a more egalitarian dialogue with those who were racially marginalized ensued. This dialogue has become a space through which it is possible to articulate deep, taken-for-granted assumptions about whiteness and thereby begin the process of liberating our sociocultural experiences from racial oppression. Conversations on whiteness have not reached fruition yet in the United States, but the point here is to illustrate the meta-theoretical argument that a dialogue inclusive of the margins was NECESSARY and that this goal of the dialogue is liberation for all of us. According to Harding (2003), marginalized perspectives should constitute the place from which research is generated because of its inherent meta-theoretical relation to the mainstream.

Moral Relations

"In the traditional view, the exclusion of moral, social and political influences from the sphere of scientific discovery and debate was necessary to its objectivity" (Benton and Craib, 2001, 50). In contrast to the traditional view, research at the margins is morally connected to emancipatory research goals through researchers' particular commitments to making the world a better place for everyone, with particular attention to those people whose lives are compromised by the status quo. Marx and Engels maintained that science had the potential to liberate humanity from disease and other problems of the social world. However, Marx (1999) also argued that it was possible for the forces of domination to be hidden in the production of scientific knowledge. Marx indicated that researchers must critically question their work and its outcomes in order to maximize the potential for liberation and limit the possibility for masking oppression. Marx astutely pointed out that research could be used to perpetuate oppressive relations if researchers did not take up a critical perspective (Benton and Craib, 2001). The atrocity of the Tuskegee experiments[1] is the story of researchers willing to harm marginalized African American males in pursuit of so-called truth. This study perpetuated the values and power relations of racial oppression (Jones, 1993). Once discovered, the study was highly criticized and legally challenged (the National Association for the Advancement of Colored People filed and won a class action suit in the 1970s). Work at the margins should heighten a researcher's moral commitment to the well-being of all people (Harding, 1991). Moral questions should provide a critical springboard from which to assess the acceptability of research outcomes for all people, particularly those who might be disadvantaged by the study.

So far, this section has explored what it meant to talk about a link between liberatory possibilities for research and doing research at the margins. Having clarified what constitutes that link factually, metatheoretically, and morally, it is now possible to move forward with a justification for the main argument of the chapter: Research at the margins is uniquely necessary for research intended to contribute toward social emancipation and liberation. The remainder of the chapter is devoted to justifying the proposal that research at the margins should be valued and supported precisely for its liberatory potential. Both practical ramifications and theoretical justifications are offered.

Practical Ramifications

This section of the chapter provides an exploration into two aspects of the research practices associated with the practical link between non-mainstream research and emancipatory possibilities: (1) the means by which research has been marginalized and (2) validity concepts involving social concerns. Brian Martin, an advocate for "liberation science," wrote, "Being a dissenter [in research] is still risky with many dissenters coming under attack," however "[o]ne nice thing about promoting alternatives is that they include built-in critiques" (Martin, 2000, 10–11).

The Means through which Research has been Marginalized

"Out of the many things scientists do, they attach meaning to only some things, which they call 'doing science' or 'applying the scientific method'" (Martin, 1992, 84). Across and within research communities there have been reasons why some research would be considered at the fray—not as legitimate or valuable, and not as important or noteworthy, as other investigations. Often times, the standards through which a research community evaluates the quality of its efforts are assumed to be inherent in the scientific procedures/methods and neutral aims of study—just good practice unconditionally. Herein lies a basic problem. Some notable scientists from the margins of their disciplines have argued, "Because all inquiries and knowledge claims occur in social contexts by persons with cognitive, emotional, interpersonal, and other commitments, biases, and ideologies, all research can, of course, be said to have a normative component" (Woodhouse et al., 2002, 298). Normative components are both the product of and the material for furthering mainstream research practices regardless of the potential social problems that might surface as a result.

Assumptions about what makes good research constitute the heart of a research community. Moreover, the pulse of a research community is established in the conjoining of economic and political forces that render some research more fundable, more publishable, and more accessible than other research. Scholarship that falls outside this heart and pulse are necessarily on the margins. Marginalized scholarship provides normative and evaluative challenges to mainstream research practices. This does not, in and of itself, result in liberation from unrecognized oppressive structures or limitations to equity, but it provides an opportunity for such. For example, researchers from the margins have recently challenged the research postulate that the researcher must be aloof and disinterested if he is to have the best chance at uncovering the facts of the situation (Korth, 2005). Feminists, among others, argued that their research actually benefits from their interested, invested position and experience. Not only that, they challenged the idea that any research position can be labeled disinterested or neutral (Korth, 2005). The mainstream perspective, easily identified by many researchers, might be interpreted as neutral because of its taken-for-granted recognizability. Yet, there is no evidence to support the claim for which a partisan researcher differs from a self-professed neutral researcher in terms of either their abilities to access the so-called facts of a situation or their willingness to learn from the research even if the findings contradict one's expectations or hopes. More important, voices from the margins contribute to the research conversations precisely because they query the taken-for-granted values of mainstream social science. This challenge carries with it the possibility that social science practice and theory might be liberated from the pretension of neutrality and pushed toward more refined concepts of objectivity, such as a concept that describes in detail a variety of third-person perspectives (Carspecken, 1999).

Another example of a marginalized set of voices in the contemporary social science conversation involves the use of statistics. In the 1970s, a group called "Radical Statistics" (Radstats) coalesced in response to a shared concern over the political implications of their work and the common misuse of statistics. Radstats members are "radical" because they "are committed to helping build a more free, democratic and egalitarian society" (http://www.radstats.org.uk/about.htm, accessed October 9, 2008). Members of Radstats are concerned about the extent to which official statistics reflect governmental rather than social purposes (http://www.radstats.org.uk/ accessed March 15, 2008). This group decries the use of technical language to exclude questioning and limit conversation. It also dismisses the idea that statistics should direct policy, arguing instead that statistics provide only one form of information that might be useful for making appropriate policy decisions. Radstats members collectively cite a concern for society as the basis for their various challenges to the typical ways of doing statistical business.

Looming just under the surface in the above discussion is the question of validity. Are the mainstream approaches to doing research in place precisely because they are more valid, or do those advocating nonmainstream approaches to research have sound validity concerns about mainstream social science?

Validity Concepts Involving Social Concerns

The British Society for Social Responsibility in Science was formed in 1969 to publicly raise questions regarding the social effects of science. Nobel Laureate Maurice Wilkins, who led that group for quite a long time, believed that scientists possessed social responsibility for the detonation of the atomic bomb in Hiroshima and Nagasaki. Wilkins had contributed to the development of that weapon. He described feeling remorse when the bomb actually worked. A disheartened Wilkins turned from physics to biology. Wilkins' primary criticism involved raising a different expectation for science, justifying it according to its positive effects—its contribution to making the world a better place.

Research is meaningful through a heterogeneous family of validity concepts that address the social and moral effects of the research findings for the communities affected and involved. In this section of the chapter, I introduce validity concepts that refer to the moral and social impact of research on local communities. These concepts are themselves marginalized according to mainstream social science discourse. The marginalized status of these validity concepts is telling and illustrative of itself.

In a previous publication, I (2002) argued for the effects of consciousness-raising as a point of validity. I argued that the validity of research is, in part, dependent upon the opportunities afforded to participants in the participation itself. Consciousness-raising is one such opportunity. This challenges the notion that research should not change or affect its participants (control them, but leave them unaffected). The validity question involved here is: To what extent does the research provide participants with opportunities to better their own lives? Freire (1977, 2000) thought that both the oppressed and the oppressing people could better their lives only in relation to one another, and this is something that research with and at the margins can/should foster.

Patti Lather (1991) developed the concept of "catalytic validity" to suggest that the validity of research rests, in part, on the activities it inspires for those who participated in the study. The question at the heart of catalytic validity is: What does the research mean for future actions and choices of the participants who engaged in the research? Traditionally, research findings have not been returned to participants and have not been held accountable to such a question, and, therefore, catalytic validity challenges mainstream traditions.

Similarly, but in a different field of research, Messick suggested that researchers ought to consider an assessment of the social consequences and functions of their studies (Fraenkel and Wallen, 2002). This way of thinking about validity acknowledges that not all research is worth conducting.

If we accept that the validity of research involves, in part, its effects on the social world, then we accept the assumption that such effects should be at least benign and at best emancipatory. Emancipatory possibilities in a society depend upon unleashing freedom and egalitarianism along the fringes of social life and through the disruption of inequity at the center. Thus, the question of social impact cannot be answered without knowledge of the margins.

The above validity concepts are typically applied to individual studies, but it would make sense to apply the same concerns to research writ large. If research communities, as a whole, are held accountable to the same validity questions of social impact and the hope for emancipatory potential, then the margins of methodological practices become equally central to claiming the worthiness of whole bodies of research. For example, randomized control trials are considered the gold standard of research in psychology. Critical qualitative research stands at the margins in this discipline (Parker, 1997). However, critical qualitative research contributes to the overall value of research in psychology precisely because it implies the question: What are the effects on the social world of a psychology only explicated through randomized control trials? We know that, in general, the answer to that question is that it excludes particular moral questions (Parker, 1997), particular life questions (Martín-Baró, 1994), and the perspectives of marginalized people.

Having examined research practices with respect to justifying the proposal that liberatory research purposes can be met through the conduct of research at the margins, it becomes important to explore plausible theoretical justifications. Providing a theoretical justification helps to ground the argument in a way that can be drawn upon by a variety of research approaches and methods.

A Theoretical Justification

Presently, the theoretical landscape of the social sciences is a hotly contested and factionalized terrain. The very idea of a grounding theory of research is eschewed by some. Moreover, others treat methodological theories as if they are a collection of outfits to select from and put on for particular occasions. Many would not accept the proposition that theory can provide an adequate justification for methodological decisions. Contrary to each of these popular views, critical theory suggests a provisional, intersubjective[2] ground where dialogue and position-taking are fundamental. We can envision this ground as a "conversation," but there is more to it than that. Criticalism is one theoretical context for which research at the margins is not only sensible but necessary in terms of both understanding social life and bettering it. The necessity for the margins can be articulated clearly through the critical theoretical principles of egalitarianism, openness, and mindfulness toward an ideal. These particular aspects of criticalism supply theoretical justification for the link between doing research at the margins and emancipatory hopes.

Egalitarianism

Congruent with democratic principles and universal discourse ethics (Habermas, 1998), egalitarian principles of critical theory demand the egalitarian inclusion of the margins (Korth, 2006). This is because, according to critical theory, all claims to truth (including relevant experiences, interests, and perspectives) must take equally into account those whose lives would be affected by the claims—with mainstream perspectives not given more weight or power. No persons/groups should be barred from expressing themselves honestly and clearly on an issue that has consequences for them (Habermas, 1981). When translated into the practice of methodology, this means that the voices and experiences of those at the margins must be included in one's research scope, aims, and methods. Theoretically speaking, empirical or methodological conversations are not complete or valid if the relevant perspectives are not included in an egalitarian way. Therefore, methods that bring the margins into the conversation in an equal way are important to the larger social science endeavor. Carol Gilligan's (1977) expansion of Kohlberg's research is a good example of this theoretical point. Her research brought the marginalized experiences of women into the "conversation" of moral development. This was necessary because Kohlberg was proposing a human theory of moral development whose claims included women, but whose sample did not.

Openness

Critical theory requires that conversations be open to dissent, in part because one's claims to truth are fallible. At its inception, dissent requires taking a "no" position on claims to truth. The "no" position is one of the roots of intersubjectivity according to Mead (1954). Taking a "yes" position leaves assumptions intact, but taking a "no" position requires an explication of assumptions. The intersubjectively structured yes/no position requires interactants to understand the assumptions and reasons that make the interaction comprehensible (Tugendhat, 1986). When a conversation is open to dissent, it produces an increased opportunity for grasping inequalities and distortions in social life, particularly when the dissent emerges through marginalized experiences and methods.

This principle can be applied to the research community by making sure the methods encourage an openness to dissent in the research conversation. In a lecture given to the Institute of Science in Society, Wilkins (1999) recalled a conversation with his colleague Crick who claimed that the secret to his successful working relationship with his fellow researcher Watson was their frank and open communication with one another. The three of them won a Nobel Prize for their work with Rosalind Franklin to discover the DNA double helix. Wilkins went on to suggest that open dialogue between scientists who do not share views might be the most important vehicle for keeping science accountable for it social effects.

The margins are relevant to the theoretical call for openness. The mere conduct of methods at the margins is an act of dissent and, thus, the margins force an openness to the research conversation. For example, Vine Deloria (1997) used the marginalized experience of Native Americans as a voice of dissent to the story of U.S. settlement, science, and white privilege. The openness of the scholarly conversation in history makes possible the liberatory effects of destabilizing the hegemony of white stories and privilege. The use of qualitative approaches in the field of counseling psychology provides another example of dissent— qualitative methodology constitutes a break from the more mainstream adherence to the randomized control trials (and the concomitant assumptions) more common in the field of psychology (Parker, 1997).

Dissent is important because it brings new things into the conversation (Martin, 2007). It supplies the opportunity to call taken-for-granted methodological assumptions into question. Social science fields have to be open to dissent if our understanding of social life is going to move us toward freer, more democratic, and less oppressive communities. In theory, democratic expectations require democratic means. The effects of such theory in the real world would include the possibility to identify the boundaries of the margins, to decenter the mainstream assumptions, to establish more realistic descriptions of the phenomena, and acknowledge the fallibility of research itself.

Mindfulness of an Ideal

Critical theory avoids radical relativism[3] by invoking an ideal (Carspecken, 1999). The ideal is considered provisional and adjustable, but not avoidable. There is a conception of the ideal that underlies all actual efforts at social life. This is not just a practical point but a theoretical one as well. The potential for understanding and for coordinating social life relies on the generative power of an ideal (Korth, 2007). Critically speaking, the content of that ideal comprises egalitarianism and openness (Korth, 2006). For criticalists, the ideal (as it stands at the moment) is freedom from oppression and inequity. Realism is accepted—that is, when researchers are able to document oppression, it is assumed that the oppression is factually depicting the real world. When these facts are disputed, the critical process is underway. This, however, does not result in an untethered competition of facts (Carspecken, 1999). In addition to the dispute of truth claims,[4] critical power is linked to the discrepancy between the facts of the real world and the values implicit to the ideal. Criticalists value a freer, more democratic, and egalitarian social life (Kincheloe and Mclaren, 1998).

Being mindful of the ideal is benefitted through methods at the margins in both practice and findings. The ideal must be applied to methodological practice—this ideal would most specifically include a set of social values that work toward a more egalitarian, mutually respectful, communicatively liberal society capable of supporting its members in their quest to be fully human (Carspecken, 1996; Freire, 1977, 2000). Through findings, researchers can better understand the experiences of subjects and their relation to critical ideals (Carspecken, 1996).

Doing research implicitly then requires that all persons be included in the ideal distribution of freedom and opportunity. This is not possible in the absence or silence of the margins. Thus, any given research community is ideally expected to explore the margins. Methods at the margins typically do draw on those ideals (Korth, 2005). When people at the margins are included in the research community, the conception of the good life and the values that emerge through the scholarship will be mindful of egalitarian ideals.

Conclusion

Those of us who work at the margins value the opportunity to contribute to making society a better place for all of us. To do so, we engage in research at the margins, contributing to the factual and liberatory potential of research and research communities, challenging research communities to practice what they hope for in social life, calling the status quo into question, and welcoming the political and economic opportunities. After examining what constitutes the link between doing research at the margins and the potential for research contributing to emancipatory goals, I have presented here practical and theoretical justifications for the argument that we NEED research at the margins if we are to pursue emancipatory goals. In the first section of the paper I described the link between marginalized and liberatory research across three aspects: factual, meta-theoretical, and moral. The practical justification draws from the factual and moral aspects. The theoretical justification draws on the meta-theoretical aspect. Taken altogether, we must conclude: "Inquiry that aspires to the name *critical* must be connected to an attempt to confront the injustice of a particular society or a sphere within society. Research thus becomes a transformative endeavor unembarrassed by the label 'political' and unafraid to consummate a relationship with an emancipatory consciousness" (Kincheloe and McLaren, 1998, 264). The necessity of the margins for such critical aims cannot be denied.

Notes

1. In 1932, public health officials began what is known as the Tuskegee Experiment. The researchers were interested in studying how syphilis progressed, was spread, and killed. Many poor African American men who had been found to have syphilis were neither told nor treated (even when treatment was available). Instead, they were provided with free meals, medical exams, and burial insurance. The study went on for forty years.
2. Intersubjectivity is variously defined across social theorists through countless pages. Perhaps any one of these definitions could work for the broad claim made here; however, for clarity sake, it might suffice to think of intersubjectivity as the activity between subjects as they come to understand one another. This activity requires being able to take the position of the other in order to see both how one's own acts might be interpreted and how the other might have intended his or her own acts. Intersubjectivity involves a shared set of normative arrangements. Intersubjectivity is not thought to be an accomplishment of interactions but rather a presupposition to them.
3. Radical relativism is the idea that there is no moral ground or position to take other than balancing a cacophony of different perspectives, each of which is equally valid: moral, authentic, and true.
4. The idea of "truth claim" is a way of conceptualizing truth as an action— one claims truth. It is an epistemology-first view of truth. Habermas (1984) has suggested that truth claims can be organized according to their validity principles as subjective, objective, and normative.

PART II

Methodology

Uncloaking Epistemologies
through Methodology

Adrea Lawrence

According to the *Oxford English Dictionary* (1989), "education" has several meanings. One is "the systematic instruction, schooling or training given to the young in preparation for the work of life...the whole course of scholastic instruction." Another is "the process of 'bringing up'...with reference to social station, kind of manners, and habits acquired, calling, or employment" (n.p.). Taken together, these definitions have a formal sensibility about them, focusing alternately on the human and social capital that one acquires, likely through schooling, for some strategic end. The emphasis on the cognitive transmission and accumulation of knowledge, the training of children in the social norms and beliefs of the society providing schooling, and the presentation of what one has learned in the formal school setting have been the primary foci of educational research from the center. Not surprisingly, research studies of education have centered on teaching and learning in the school setting, buttressing assumptions that "education" is tantamount to "schooling." Over the past century, much educational research has concentrated on how schools as institutions have functioned over time, how most children learn in the classroom, how teachers learn their craft, and the like. Determining what is "normal" and developing ways to replicate studies and effective programs have correspondingly served as the purposes for conducting educational research from the center. Without doubt, we have learned from this. Studying education from the center, though, can limit the researcher's purview to the institutional structures of school and what is taken to be the generalizable norm. Likewise, in confining how we define education, we have, perhaps unwittingly, confined how we study education. That is to say, the methodologies we use emerge also from the center simultaneously with our center-focused definition of education. In this section, the authors purposefully consider this point and in response develop methodological tools and practices that come from the margins. Each author chooses not to import commonly accepted research practices and standards to their work in the margins. Accordingly, the authors provide a broader, more fluid definition of education that is located both inside schooling and within the larger society.

Studying "education" from the margins expands our capacity to do research in several ways: we encounter constructs that have been overlooked or simply not recognized; we encounter evidence that suggests "education" is an epic process that perpetuates and adapts cultures; we encounter groups who neither look nor act like those who are the White, Euro-American norm; and, it frees us from concentrating myopically on how the individual accumulates information, applies skills, and adopts normative positions at school. Certainly, the study of "education," or that which involves teaching and learning in the very broadest sense, is an iterative process that negotiates the often ambiguous space between the multiple poles of the margins and the center.

Studying education from the peripheries exposes epistemic rifts in theoretical positions and methodological approaches that exist between the center and the margins, revealing investigative paths that have been previously cloaked by our deeply internalized ways of knowing education as only schooling. The chapters in this section can be viewed as responses to calls in the mid-twentieth century to consider education as broad, formative processes of learning (Bailyn, 1960; Cremin, 1965; Storr, 1961, 1976). In studying education from the margins, the authors in this section highlight the relationships between individual people, groups, and their physical environments, illuminating the labyrinthine and nuanced ways in which people experience the world, learn from those experiences, and also reflect upon them. In introducing us to new epistemological and methodological approaches to studying education, the authors cast familiar ways of seeing, reading, and studying education in marked relief. By asking: How do we re-learn to see and interpret education and educational research, the authors in this section explore the themes of context, experience, and polyvocality, offering insights into how these themes can be explored in qualitative, historical, and quantitative educational research. Joshua Hunter explores how people learn through and from place; Donald Warren examines how American Indian groups have learned across generations; Amaury Nora focuses on how the researcher knows what to ask and how to interpret data when studying Latinos in higher education; and Rachelle

Winkle-Wagner considers the complexities of validating research across racial lines in ways that emphasize the voices and experiences of participants. In their topical discussions, each author closely examines what constitutes valid and reliable data or evidence, reconstituting educational research from the margins.

Context, experience, and polyvocality are methodological and descriptive terms that allow the researcher to explore actors' interactions with their physical, sociocultural, and organizational environments. Historian John Tosh (2002) notes that context, for historians, includes:

> the resources of language...the identity and background of the author, the conditions of production of texts, the intended readership, the cultural attitudes of the time, and the social relations that enveloped the writer and readers. (195)

While this understanding of context certainly applies to much historical scholarship, it privileges documentary evidence. How, then, does the researcher study and weave context for groups of people that no longer exist? In the absence of documents, what becomes the text?

Donald Warren and Joshua Hunter demonstrate in their chapters that place—the physical environment of a locale—is a text for study and often an always already contextual participant. Warren argues that the art, architecture, and agricultural practices are all cultural and physical remnants that can literally be uncovered and read through archaeological investigation and then triangulated with ethnological and linguistic study together with written accounts from early European interlopers.[1] Hunter argues that place is, in fact, a contextual participant, for humans interpret their lived experiences and shared memories through their perceptions of the places in which they occurred. This act of interpretation creates a relationship between people and place with sociocultural and physical roots. Place can also disclose events and processes that have occurred in the geologic past and more recent human recollections. The situatedness of context also extends to the present.

Amaury Nora argues that education scholars' lack of attention to the unique contexts of Latino undergraduates in quantitative studies has resulted in errors of interpretation. In his chapter, Nora writes that researchers have created their own categories of analysis rather than relying on those that might be derived from study participants. This is due in part to the exclusive use of quantitative research designs, which do not necessarily allow for qualitative means of exploring how study participants interpret the socio-academic and cultural worlds around them. Neither does a reliance on hypothesis-testing oriented around a single variable reflect the complex social and educational patterns of Latinos. To remedy these problems, Nora proposes pairing specific qualitative and quantitative research methods that examine multiple variables together and build the research design from participants' contextualized experiences.

Finally, Winkle-Wagner discusses validation in cross-racial research. She contends that validation techniques, if they at all consider cross-cultural concerns, often come from research steeped in colonialist, White-centric definitions of the "Other." How do researchers studying racial and ethnic groups, which are unlike their own, learn the epistemic assumptions and contexts of participants? Winkle-Wagner demonstrates a means for validation through which the researcher undergoes an educative process of learning from participants. Using participant-driven research methods characterized by the development of new, inductive validation techniques, she stresses polyvocality, or the many voices of participants in concert with her own voice as the researcher.

For the authors in this section, experience, then, or how we make sense of it to ourselves and others, is foundational to learning—to education—and to the methods that one uses in studying education. And, experience is not something that can be statically pinned to one individual or group. As anthropologist Edward M. Bruner (1986) writes, "Life consists of retellings" (12). This suggests that each retelling or performance of an experience is dynamic; it is also culturally grounded (Turner, 1986). As Greg Sarris (1993), a literary and anthropological theorist, notes, "One party may write a story, but one party's story is no more the whole story than a cup of water is the river" (40). Experience when recounted by one person to another is necessarily polyvocal in its interpretation, for each person's and each group's position is subjectively unique. That is, individuals and groups construct and understand themselves in relation to others and their environments, and they tailor retellings for a particular audience and setting. This makes it difficult to normalize experience across different groups of people. Warren writes that he attempts to understand different accounts of histories and memories in his study of the Arikara in North Dakota by paying close attention to the recounted and shared memories of experiences the narrator underscores, comparing them with archaeological and linguistic evidence to trace the histories in time and space. Winkle-Wagner uses the experiences of her participants as a way to begin to validate research findings in cross-racial research. Likewise, Nora explains that emically grasping how Latinos make sense of their experiences through qualitative means better contextualizes participants' experiences and allows for multiple voices to inform research design, culminating in more valid and accurately interpreted quantitative results.

Educational research from the margins is research that recognizes participants' experiences, voices, and contexts are in flux, which makes learning and teaching seem like moving targets that can never be understood with precision. What is simultaneously jarring and exhilarating is the possibility that epistemic rifts between the center and the margins do, in fact, uncloak ways in which the researcher can create paths between the two. Each of the authors in this section has done just that, providing ways of understanding education from the margins as dynamic processes for both researchers and participants.

Note

1. Curiously, understanding how to read and interpret ancient peoples places relies on research technologies that have been only recently become available.

CHAPTER 5

Researching as if Place Mattered: Toward a Methodology of Emplacement

Joshua Hunter

Introduction

The red frocks of cardinals are not easily missed against the grey lamentations of midwinter. Not unlike a couples party, the males and females split up, with males hanging around together on branches and females working the forest floor amidst transient juncos. Flocks of finches huddle in woody shrubs, only to be scattered by diligent blue jays, like pieces of blue sky come to earth. Nuthatches dance their inverted acrobatics down trunks of trees while the chickadees tip their black caps for food. All are puffed up and bulging to deflect the wind, like an overdressed preschooler who has set out in the winter world. Sunlight finds the lower shrubs full to capacity with a continually shifting citizenry, a perpetual jockeying for branches and seeds, all against the backdrop of a brooding sharp-shinned hawk loitering in the higher reaches of a poplar and the drumming of woodpeckers. Only when the hawk descends does the action quiet as all but one break for cover. As quickly as it occurred, with the hawk perched just above, tossing feathers, the scene rejoins. There's a soft blanket of snow this morning, dusting the trees and valley as it stretches upward toward the far ridgeline; the scene an iterative, veritable hive of movement and noise.

Are you wondering if perhaps you have strayed into the wrong book, a nature study book at that? What in the world is a passage like this doing in a book on research methodologies? How can this description of a natural scene help us as researchers as we attempt to unravel complex systems of human meaning and understanding? The agrarian philosopher Wendell Berry, who calls himself a "placed" person, writes that "if you don't know where you are, you don't know who you are" (Stegner, 1995, 1–3). This same observation can be applied to researchers and research—if you don't know where you research, you don't know who you research. Can it be that a researcher can likewise be placed? Place itself holds us as very few things can, and while the description of place, relegated to the status of site, is consistently sketched in various research, it too often fails to evoke either place or the meaning of place in sufficient detail.

After all, place is perhaps one of the most basic constants of human experience, yet a constant often forgotten about by our educational and research endeavors. Poet Gary Snyder (1995) asserts that

> of all the memberships we identify ourselves by (racial, ethnic, sexual, national, class, age, religious, occupational), the one that is most forgotten, and that has the greatest potential for healing, is place. We must learn to know, love, and join our place even more than we love our own ideas. People who can agree that they share a commitment to the landscape/cityscape—even if they are otherwise locked in struggle with each other—have at least one deep thing to share. (5)

Place, then, is so central that perhaps we have taken it for granted, failing to understand the significance of our relationship to where we dwell. As the scene described in the first paragraph attempts to demonstrate, research, when taking the relationship between place and person seriously, must not just describe but evoke that relationship. In this chapter, using research conducted in a Midwestern State Park, I will demonstrate this evoking process and discuss what it means for a researcher to become emplaced.

As I began research with the interpretive naturalists providing place-based ecological and cultural education at the park, one of the things I learned is that I could not possibly describe and examine the meanings of sense of place if I did not develop such a sense myself. To become emplaced as a researcher was one attempt to understand how people dwell in a particular portion of the earth, what they make of that landscape, and how they derive meaning from it. I also learned that I would not be representing the authenticity of how it is that people dwell in place if I did not account for the place itself. And by doing so, I might be able to decenter the human in such a way as to alter our perception of humanity, to borrow Aldo Leopold's (1968) words, "from conqueror of the land-community to plain member and citizen of it" (204).

Marginalizing Place

Emplacement for a researcher can be seen through the same lens as it is for an immigrant or transplant, something that has become synonymous to modernity, in which the transient nature of our society and academic endeavors is evident in our extreme dislocation from place. Nabhan (1994) suggests that:

> it is a crime of deception—convincing people that their own visceral experience of the world hardly matters, and that pre-digested images hold more truth than the simplest time-tried oral tradition. We need to turn to learning about the land by being on the land, or better by being in the thick of it. This is the best way we can stay in touch with the fates of its creatures, its indigenous cultures, its earthbound wisdom. This is the best way we can be in touch with ourselves. (106–107)

While Nabhan is more explicitly describing education writ large, his indictment of the ways in which we validate knowledge can be applied to research, in which researchers can become place-literate by being emplaced, thereby being better able to account for how it is that people dwell.

There is a larger purpose for what I am describing, and this has to do with how seldom academic research takes account of place other than in the mere description of site. The anthropological literature (and thus ethnography) is surprisingly sparse when looking at how people actively sense place, even though, as Basso (1996) and Casey (1996, 1997) both assert, this relationship is among the most fundamental to human experience. Historically, paying little attention to the basic human experience of sensing place, the intertwining of heart and mind when people come to actually dwell or inhabit a place (Basso, 1996; Casey, 1997; Feld and Basso, 1996; Geertz, 1996), has left a vacancy in how anthropology as a discipline conceives of the human/nature interface and the multifaceted ways natives construct localities and perceive and experience place. When sense of place has been dealt with, and this has been done in wonderfully evocative ways, typically the discourse portrays the topic in regard to indigenous groups, seldom examining dominant groups in society (Geertz, 1996).

Much has been written concerning how the modern experience of Enlightenment, industrialization, and urbanization have led to human societies in the West being dislocated from nature and asserting anthropocentric dominance over nature (Bowers, 1993a, 1993b; Casey, 1993, 1996; Marx, 1964; Olson, 1995; Schama, 1995; Streibel, 1999).

Cartesian duality of nature and culture, body and mind, provide a rationality for this dislocation, and the subsequent clearing of forest, draining of swamps, and alienation of emplaced people (indigenous groups) and emplaced nonhuman life make this distinction manifest.

This marginalization of place within research is a failure of the academic imagination, for not only does this limit the understanding of emplaced peoples but further limits our collective abilities to transcend the imposition of Enlightenment duality of nature and culture. The marginalization of place as site leaves it homogenized, institutionalized, and stripped of any intrinsic worth, something held apart from human culture. The reality is that place has provided the grounding of our understanding of the earth and of ourselves as emplaced beings. What is evident from research concerning people and place (Basso's work being one among the best) and my own research is that there becomes apparent a knitting together of self and place, an intertwining of landscape and human identity, spectacularly refuting the nature/culture divide. To take this a step further then, the preservation of place and of humanity as emplaced beings is inextricably bound, and any research intent on fostering preservation should dwell deeply upon this. This fosters a better understanding of the interdependence of nature and culture, of the natural and human worlds.

Marginalizing the nature/culture interface has diminished our understanding of the power of these relationships in evoking whole worlds of meaning for people and communities. What is ultimately perpetrated is a continued anthropocentric worldview that undermines ecological understandings and a broadening of our notions of community to include that which is not human. Basso's (1996) critique of research, which is purely materialist in orientation in its analysis of place, renders the writing sterile and unidimensional. This is why, as Basso (1996) claims, poets and journalists are better able to apprehend and evoke place and its meanings than are researchers. Place, then, is marginalized in research methods and in how it is described in research writing, both of which fail to recognize its great import for how it is that people live. Being emplaced, as described in my own research, is a repudiation of this marginalization, a repudiation that validates the connections to an area and to the community of life that surrounds us as humans.

To counter this historic trend, I offer this example of privileging place as a central character of the study, one imbued with its own subjectivity, one worthy of our care and attention, and to critique persisting anthropocentrism and Enlightenment dichotomy between nature and culture. In the end, I hope that this offers a perspective on research that broadens our conceptualization of community to include nonhuman nature, which can only have positive results for preserving that which is going under fast (both natural systems and emplaced ways of knowing), where it has not gone down already.

What, then, do we make of this reciprocity between people and place? This relationship of meaning-making inherent between people and place becomes iterative and dynamic, for as we perceive or sense place, we ourselves and also our senses become emplaced, forming a dialectic between ourselves and place. This is the intimacy born between humans and place that encourages a reciprocity of meaning-making and of the lived body and place. Casey (1993, 1996, 1997) suggests that since we are never without perception, the dialectic between perception and place signifies that we are never without experiences that are emplaced, thus "we are not only *in* places *but* of them [his emphasis]" (19). This is the significance of what Basso (1996) calls "interanimation," and is of great similitude to what Michael Jackson (1998) describes as the "intersubjective turn" within anthropology, in which it is imperative that scientific rationality not displace "those anthropomorphic correspondences that enable people, in moments of crisis, to cross between human and extrahuman worlds" (6).

With all this in mind, we need to get back to the work at hand, back to what Casey (1996) suggests will bridge the pre- and postmodern worlds, principally coming back to place, in which place and human perceptions of place were and can be again primary. As researchers attempting to understand the human creature as an emplaced creature and as a vehicle for authentically evoking place in our work, this can only have positive results. But what could this possibly involve?

On Being Place-Literate

Researchers must be place-literate. This point may be so obvious that it could go without saying, but just as place is too often taken for granted, so are some of the fundamental aspects of research methodologies. Researchers must be place-literate, and this requires a significant amount of effort to understand place for its intrinsic qualities and subtle arrangements. What this means, of course, is getting out and learning the lay of the land for yourself. This can only mean getting out and walking and sensuously experiencing place with all senses functioning: walking forest paths; walking neighborhoods; walking school hallways; walking those conduits of human experience, memory, and community.

Walking by yourself; walking with a native or a local expert; walking with groups of people as a way to gauge their varying reactions; walking with field guides, city guides, or books of poetry, and learning that the place itself brings out its complexities, grandiose vistas, and its nuanced whispers.

This requires spending time with place as an informant to your research, as a source of wisdom and insight that may engender information about human informants. What is fundamental to this is the actual experiencing of place sensuously, emotionally, and cognitively as a native would, to comprehend the great scope of ways people relate to the land. To gain the insider perspective, to reach into the emic way of knowing, to access the native worldview, one must begin to comprehend place as a native would. Barry Lopez (1990), in his book "The Rediscovery of North America," suggests that:

> When we enter the landscape to learn something, we are obligated, I think, to pay attention rather than constantly to pose questions. To approach the land as we would a person, by opening an intelligent conversation. And to stay in one place, to make of that one, long observation a fully dilated experience. We will always be rewarded if we give the land credit for more than we imagine, and if we imagine it as being more complex even than language. (37)

What this further allows is an understanding of the nonhuman life and landscape in a deeper way. This can only benefit the researcher who is intent on understanding how it is that the people relate with nonhuman life and dwell in the landscape.

Thus, a researcher must begin the process of developing a sense of place if she is intent on understanding the complexities of the human/place dyad. But what does it mean to sense place? Sensing place is dependent upon a fundamental understanding of the ecological, historical, and cultural structures that give a locale shape; and to do so by actively experiencing place sensuously, cognitively, and emotionally. To better understand these structures, I am going to borrow a term from historical analysis and suggest that one must come to understand le longue duree of place.

Now, le longue duree as a heuristic device gave priority to long-term historical structures over events. The approach incorporates social scientific methods into history and was pioneered by Marc Bloch and Lucien Febvre in the first half of the twentieth century. Appropriating this method into understanding place requires knowledge of the natural and cultural structures by which a place, as a vibrant and unique locality on the globe, came to be.

The Emplaced Researcher: Theory and Exemplars

To illustrate, the location of my research requires me to understand the geologic and cultural historic structures of limestone if I am to understand the ways in which people interpret, explain, and develop a relationship with this product of both geologic and human constructions. The land upon which my research takes place was once part of a great inland sea, called by poets the Sundance Sea, and the rock upon which various human actors stand was at one time ocean floor teaming with precambrian life. Only through the geologic process of sedimentation did the mud harden, entombing millions of creatures in the various layers. This sedimentary rock, called limestone, has formed great bands over the millennium, each one signifying a grand sweep of geologic time.

Limestone also has the capacity for eroding away until it becomes a honeycomb of caverns, sinkholes, underground rivers, and passages, which geologists call Karst topography, and which is important ecologically as a unique habitat, historically for its use by various human groups, and culturally for its ability to lure the fascination of humans to view it up close and to encounter the stone sensuously and intimately. There is another part of this story of limestone that entails the great digging out of massive stone blocks to construct buildings both locally and in great cities far removed. Quarries opened up the earth to dig and blast out the most symmetrically square blocks, with tight pores and fine grades. These came from the older layers of limestone, and if gazed at with a magnifying glass, even in a building such as the U.S. Capital, you could see the small shells or bones of tiny crinoids or brachiopods. It's a great bit of irony that our own colossal buildings and monuments are made out of this congregation of mud and the remains of tiny bodies of long gone animals and plants. And it is evident that this one type of stone can account for an incredible stretch of geologic, ecologic, and cultural history. Le longue duree of place requires an understanding of these structures, for they provide the people who work, visit, and love this place with meaning in a myriad of contexts. As one of my informants, Maggie, told me, "the story of this place is geology. It all begins there."

Writers within the genre of sense of place typically employ a phenomenological lens by which to understand how place is perceived and in what ways people come to understand their own awareness of place and themselves (Casey, 1993, 1996, 1997; Jackson, 1998). Merleau-Ponty's (1962) insight that to be in place is to perceive, to become aware of one's consciousness and sensuous presence in the world, elicits not only an active sensing of place but of ourselves as well. Perception, in this regard, becomes the groundswell of our understanding of place and of ourselves as emplaced creatures.

In fact, if we are to assume that perception of place is central to our own subsequent understanding, then phenomenology is an appropriate starting point. For within the anti-enlightenment stance of the phenomenological approach, specifically addressed by Merleau-Ponty (1962), Heidegger (1977), and Casey (1993, 1996, 1997), perception is the primary locus of local knowledge, intrinsically experiential as "lived experience" or *erlebnis*. As we perceive or sense place, we ourselves become emplaced, forming a dialectic between ourselves and place. This is the intimacy born between human and place, which encourages a reciprocity of meaning-making, and of the lived body and place. This is further the intimacy that a researcher, sensuously aware, should strive for to better capture the complex relationships between people and place.

Heidegger's (1977) notions of being in the world, his Dasein, or dwelling in the world is of importance here, for "dwelling is said to consist in the multiple 'lived relationships' that people maintain with places" (Basso, 1996, 106). These lived relationships are often reciprocal in that people derive meaning while dwelling in place, yet they also inscribe place with meaning, thus intertwining their own identity with the place they come to inhabit. A sense of place could allow people to more directly experience place and thereby allow place to be more fully brought into being. As Casey (1996) asserts, "we come to the world—we come into it and keep returning to it—as already placed there. Places are not added to sensations any more than they are imposed on spaces. Both sensations and spaces are themselves emplaced from the very first moment, and at every subsequent moment as well" (18). We cannot examine the understanding of an individual or of cultures without referencing how it is that people dwell. This basic component of human existence, often marginalized within research, can afford the researcher open to experiencing place a firm grounding in the age-old dialogue between people and place.

The experience of dwelling and the validation of this knowledge-base serve as a critique of Enlightenment discourse on the primacy of reason at the expense of experience. Gadamer (1989) asserts that it is this Enlightenment emphasis on reason and the importance of self-consciousness that served to distance humans from experiences that enable us to understand our own existence, a condition he calls "alienation." I would assert that researchers can similarly fall victim to this condition by invalidating our experiences in the field, perhaps with the field, as the case may be, and thereby distancing ourselves from understanding more fully what we are encountering. This again, is an instance of marginalizing both place and emplaced experiences, and can further dislocate the researcher from the meaning-making between people and place.

Now back to the birds, and an accounting of how experiencing bird song can inform the researcher of the value of emplaced experiences. Seeing the birds, hearing them, following their calls, noticing their nests and marks, is tantamount to developing appreciation for the unique relationship between people and birds.

To hear a wood thrush's lilting song in evening woods is indeed magical, and knowing and experiencing this as a researcher is a first step in understanding why it is that people get so caught up in trying to hear that song. If a researcher were intent on understanding the complexities of communicative acts between people, she would invest herself in knowing nuances of language and various intentions of the speech act. The same attention is owed to the relationship between humans and birds, or other animals, or trees, flowers, even stones for that matter. People travel thousands of miles, often following the migrations of winged creatures, track species over the course of a lifetime, marvel at the songs of different birds, to the extent that they learn the diverse calls of each species, are able to recognize those calls over the cacophony of the forest, and learn to sing the songs themselves to bring the birds in. I have, on numerous occasions, come into the nature center office to hear turkeys, owls, cardinals, crows, phoebes, and others singing and calling. I was not surprised to find that no birds were present, only a group of naturalists feeling the avian urge to sing: practicing for the public, sharing knowledge about different calls and species, and sharing their melodic talents with each other. Anyone who has been in the midst of a group of birders cannot be but impressed by the complex relationship and meaning-making taking place.

A naturalist in my study, named Joan, described an experience to me that can better illustrate the importance of a researcher being emplaced. Joan described her first hike at night with a group in the park, in which she intended to highlight the changing natural scene as day faded into night and a new cast of characters wake to the nocturnal world. Her intentions were interrupted by the group getting lost, and this troubled her a great deal as she had no clue where she was. Having hiked that area of the park, I was familiar with the intricate weaving of trails along the crest of those steep canyon walls. And because of this, I better understood her trepidation and anxiety. She felt that the hike was ruined. At this point Joan stopped talking, and a smile grew on her face, for she then explained how it was at this point that she, for lack of something else to do, began calling for the wood thrush, and the wood thrush answered. The wood thrush is migratory and shows up in the early summer, and sings in the evening and early night hours a song, like a magic flute player. Joan talked about how excited the group became, not only in the beauty of the song, but in the fact that they were singing to each other, two dissimilar musicians singing along the same forest path. Joan said it was one of the best hikes she has ever led; the magic intermingling of human, bird, night, and place in that one moment. It's hard to understand why people get so caught up in this type of experience until you've had the pleasure of hearing the wood thrush sing, and, if you're lucky, singing back and forth. Emplacement for a researcher is a tool of understanding, which allows for a more empathetic approach and yields a better sense of the other's perspective.

Here is a final example of the importance of developing a sense of place by a researcher. While observing a natural history program highlighting coyotes, I documented an interaction between an interpretive naturalist and a park visitor in which the visitor was enquiring about the appropriate times for shooting the animals. She explained that she saw them outside her farm house, that they weren't hurting anyone, but that she had young horses and wanted to know if she could shoot the coyotes for simply being on her land. She said she had a .45 pistol and wanted to go out on her front porch and start firing. She asked if this was against any laws or was problematic when there seemed to be so many coyotes. The naturalist replied with an ecologically sensitive response about the importance of coyotes as predators and that if the coyote was not harming anything, then it would be wrong to begin firing at it. The woman visitor shrugged at this suggestion by remarking that it would be better to shoot them now than wait for something to happen later.

From where does her animosity arise? Is it personal? Historical? Cultural? Does this relate to her experience of dwelling in this rural place, in which coyotes seem to procreate with wild abandon? This dialogue is a powerful example of the confluence of both natural history and human cultural history. Nature and culture cannot be separated from the other, and in this light her comments must be interpreted through a cultural lens of symbolic interpretation, requiring an understanding of an ecological and historical legacy of animosity against these creatures and of a culturally symbolic relationship between humans and other creatures, specifically predators. The central question, then, as a researcher is, how can I understand this woman's animosity for the coyote if I cannot account for the animal itself? Its natural history is tied to that of our own (Euro-American, that is), its expansion is tied to our own expansion as we cleared forests and exterminated larger predators, its ability to adapt so closely to humans mirrors our own species' abilities and social networks. The historical and persistent extermination of these animals, along with other large predators, must be understood ecologically, culturally, and symbolically to begin to extrapolate this one woman's wish to, as cliché as it sounds, "blow away these varmints."

Conclusion

When things are constantly being related back to place or its inhabitants, it becomes incumbent upon the researcher to begin a process of emplacement to more fully account for the human/place dyad. But understanding must be rooted in historical, cultural, and ecological processes, le longue duree of place, as one way of gleaning a deeper sense, a grounded sense, of lived reality. The methods, then, can reflect the theoretical underpinnings of phenomenology, and research on the ground can demonstrate the importance of being emplaced. A fundamental critique that Basso (1996) levies against cultural ecology is that it relies too heavily upon a materialist examination of human/place interactions, rather than delving into the symbolic and interpretive ways individuals and communities relate to place. Delving into these symbolic relationships suggest an intimacy of knowledge being generated by the researcher. Intimacy of knowledge requires an intimacy of experience, a sensuous scholarship, as Paul Stoller (1997) suggests, that brings the researcher into direct experience with phenomena and place itself through sensuously relating to experience. Our awareness is dependent on this, and while most research has marginalized both place and the sensuous experience of place, it is only to the detriment of our collective understanding of how it is that people dwell. To discern meaning-making people have with the basic element of emplacement, the researcher needs to be emplaced herself, only then can a more truthful representation be made. This process is also about being open to the expressive and emotive ways that place is felt, imagined, perceived, and remembered. This moves beyond a purely materialist orientation about how humans use the natural world, expanding our knowledge of what people think, feel, know, and remember of their own presence in the natural world. It is in this way that scholarship can be more authentic, creative, and expressive.

CHAPTER 6

Twisted Time:
The Educative Chronologies of
American Indian History

Donald Warren

Although countless public celebrations around the world announce otherwise, Christopher Columbus was not the first outsider, certainly not the first European, to set foot on what became known as the Americas (see Axtell, 2001; Mann, 2005; McGrath, 2007). He and his predecessors, transient sailors, traders, adventurers, and, in time, occupiers and conquerors, found diverse climates and topographies bearing clear signs of human manipulation. They were not discoverers, for the lands were inhabited. Whether nomadic or settled, the indigenous peoples varied as widely as their natural environments. They knew how to manage Arctic tundra, deserts, forests, animal populations, and farms. They shared a religious awe for nature's power and unpredictability, but expressed spirituality in distinct theologies, symbols, and ceremonies. Unlike their counterparts in Europe and the Mideast, there is little to no evidence that they engaged in religious persecutions among themselves or fought religious wars, although some, like contemporaries on the other continents, practiced human sacrifice to deities. Their societies interacted with each other culturally and commercially, whether peacefully or militarily, exchanging goods, arts, technologies, seeds, rituals, languages, and ideas, leaving mixed accumulations of relics and garbage for others to ponder and identify. Some established vast empires that rivaled those elsewhere in the world relative to population size, urban density, and cultural sophistication. Most of those living north of the Rio Grande River did not have written languages but fashioned other means of marking events and the passage of time; their histories are thus encrypted within oral traditions, calendar sticks, and other mnemonic devices. In general these were peoples with extended and routinely exercised memories. Their legends and totems depicted societies armed with strategies and beliefs for anticipating change, bending it or adapting to it, and enduring calamity. Some succeeded, across tumultuous millennia, although none was fully prepared for the shocks that came after 1492. Most non-Indians are unfamiliar with these other ancient regimes.

What follows is a progress report on segments of American Indian history (see Mann, 2005; National Museum of the American Indian, 2007; Trigger and Washburn, 1996). It emphasizes ways the indigenous peoples of North America maintained, grew, and, when stressed or inspired, amended their societies prior to European contact. Confronted by insurmountable hardships, some moved to new lands, altering practices and institutions accordingly. In short, the chapter offers episodes in the history of education (Sorr, 1961). It draws from research sources and methods that together have provided prisms for refracting light from Indian perspectives across elusive pasts, portions of which remain, for now, otherwise unrecoverable. Recently discovered documents, new archaeological findings, and fresh conceptual contributions, among other advances, portend a state of knowledge in motion. For the foreseeable future, it is likely to remain fluid, dynamic, possibly wrong, and preliminary, conditions analogous to education itself. Such caveats have been posted in the historiography of slavery in the United States, where sources and methods continue to upend one-sided claims that slaves failed to exert agency, or that owners and whites generally paid only minimally in terms of lost lives, opportunities, and social capital for the institution of human bondage and the racial "color line" bequeathed to Americans who came later (see Du Bois, 2003; Hahn, 2003; Warren, 2005). On the topic of slavery, as with American Indian history, the angles cast by different perspectives change the subject and reorder prospects for achieving reliable coherent narratives. In both cases education emerges as an explanatory process (see Corle, 1972; Jones, 2003; Blackmon, 2008). But the substantive and methodological details vary sharply, as this chapter's focus on selected Indian societies intends to illustrate. Slavery as an educational institution requires other unique approaches, methods, and sources, and its own chapter.

Historiographical Foregrounding

The history of American Indians has not yet received the full attention of education historians, despite efforts by Adams (1995), Deloria Jr. (1991b),

Lawrence (2006), and a handful of other scholars (see Lomawaima and McCarty, 2006; Monaghan, 2005; Mihesuah, 1998; Swisher, 1998). Emphasizing nineteenth- and early twentieth-century sections of admittedly under-explored historical territory, these authors have begun laying a theoretical foundation. Not surprisingly, their accounts focus on the imposition of Euro-American forms of schooling on native peoples. For one reason, primary sources, government reports, and census records are readily available in print and electronic archives. Although buried in often obscure collections, thus requiring labor intensive investigations, the trail remains accessible if not warm. Recent contributors to this work tend to cast Indian exposure to white schooling in the context of nineteenth-century proclamations of manifest destiny. Research dwells on the explicit intentions of Anglos and Spanish to save Indians by destroying their cultures, classifying the project as an instance of a hydra-headed and efficiently hegemonic colonialism. Ostensibly to redeem native souls through conversion to some branch of Christianity, Adams, Deloria, and Lawrence (2004) argue, the aim is more accurately portrayed as a determined quest by outsiders for profits and Indian lands. Slaughter or, short of extreme measures of extermination, forced relocations became acceptable strategies. Both were followed throughout the country from the seventeenth century onward, north, south, and across western territories, as treatments of the Pequots, Cherokees, and Dineh (Navajos) illustrated (Axtell, 2001). Education policy entered the plans well before Thomas Jefferson's day, who had much to say on the subject, and it too followed the colonial pattern (Wallace, 1999; Ellis, 2007; Jefferson, 2002). Taking Indian children often to distant schools, separated from their parents and communities, forbidden to speak their languages or practice ancient rites, dressed in Anglo garb, and shorn to look "American," seemed to whites at least a humane alternative to overt brutality (Lomawaima and McCarty, 2006; Reyhner and Eder, 2004; Szasz, 1988; Wilson, 1998). Most Indians failed to grasp a meaningful distinction between the two strategies. The analysis has opened doors to other inquiries. These authors want to examine what Indians risked losing along the way toward schooling. Yes, many eventually learned to speak and write in English and acquired other basic cultural and economic tools of assimilation, but the costs proved dear to Indians and non-Indians alike, notably the loss of native tongues, and with them the oral traditions that had been maintained through centuries-old routine rehearsals of the capacity to remember (Monaghan, 2005). In danger as well were a unique spirituality and most fundamentally the cultures that freighted and organized communal understandings for individual tribal members. The goal was a frontal assault on Indianness.

The schooling of American Indians has become a frequently visited topic in the history of education, as even a cursory review of survey textbooks reveals (Rury, 2005; Spring, 2008; Urban and Wagoner, 2004). By contrast, the field offers comparatively less that advances the other research purpose: inquiry on the educational relevance of the cultural histories of Native Americans, their determination to protect and enrich these reservoirs of meaning, and their persistent assertions of independence.

This chapter posits these larger, multifaceted, and contested cultural dynamics squarely within the purview of historians interested in education. So understood, American Indian history becomes an epic struggle over learning, where and how it occurred, and what costs and benefits followed in its wake. The approach poses methodological difficulties with substantive implications. Different methods and theoretical constructs capture different data. The analyses veer accordingly; different stories emerge, as the chapter attempts to demonstrate. Sources in familiar documentary formats that mirror Indians' vantage points are not readily available. Relevant materials lie embedded in the specialized literatures of anthropology, archaeology, botany, economics, and geography, among other disciplines, and some can only be found farther afield in the surviving stories, myths, and memories of Indians themselves. These are not the usual sources consulted, particularly in concert, by education historians. To ensure that Indian perspectives are weighed, this culturally attentive history requires abilities to "read" and contextualize oral traditions, art, and artifacts; religious beliefs, symbols, and performances; and institutionalized practices that bear little similarity to Euro-American concepts, conventions, and social organizations. The relevant literature surfaces as a blend of ancient sources, some of which have been translated into English only in recent decades, and research findings in multiple disciplines that have accumulated over the nineteenth and twentieth centuries. Not all of the latter bear scrutiny, given the overt and sometimes subtle taints of racial stereotyping and faulty science. The chronology is twisted, looping back and forth across time, leaving a contorted trajectory rife with educative potential. We can learn from it.

Historians of the United States confront similar difficulties, in part because of their own history. Much early study of native America rested uncritically on outsiders' perspectives and sources, producing Indian history without Indian testimony. When history emerged as an academic discipline at European and American universities in the nineteenth century, newly professionalized historians recognized early on that the indigenous peoples of North America, and by extension natives every-where, posed significant, perhaps even unique, methodological and substantive dilemmas (Conn, 2004; Galloway, 1997). The science of history required documentary evidence, but at the time most of the sources accepted as reliable had been produced by non-Indians. Almost uniformly, these materials echoed commonly held (white) views on native peoples, namely, that they were uncivilized, irreligious, unedu-cated, and brutal, savages in several meanings of the word. They were hunters and gatherers who roamed as needed, lacked cultural roots, and were guided by superstitions, and not reason, morality, and Christianity. With this slanted historiographical tradition in mind, Vine Deloria Jr. (1991a), among other native scholars, doubted whether outsiders could master the basic challenges of writing American Indian histories (see Deloria, 2003[1973]; Mihesuah, 1998; Swisher, 1998). For one thing, the perspective they bring to sources tends to be singular and unilateral, as against interactive and multidimensional. For another, they fail to consider sources Indians rely on, thus evincing grave difficulty concep-tualizing multiple narratives and competing value systems without arranging them hierarchically. Ironically, by admitting at the outset that Indians are not interchangeable, that they have always been diverse, socially, culturally, and axiologically, outsider historians could perhaps learn from research on Indians how to design and conduct more nuanced, hence more probing, investigations, whatever the subject.

Hovering as background of historical research on American Indians are often-repeated interpretive alternatives between assimilation and extinction. Although rarely welcomed by whites and the state and fed-eral governments, and denied full citizenship status until the twentieth century, Indians, it was said, should learn to function within the United States, or sooner or later they would vanish (see, e.g., Tocqueville, 2000). The dichotomy tacitly assumes that Indians (pre- and post-contact) had never confronted ecological, military, economic, or political shocks requiring cultural or organizational adaptations, that they had persisted (or not) through seasonal, repetitious, reminiscent cycles of change. They were frozen in time, like exhibits in museums of old. This per-spective implies that the indigenous peoples of the United States, with little sense of accumulated experience and only imprecise tools to record it, can be understood as lacking histories of their own prior to European contact. To apply a commonly used archaeological typology, they existed in a prehistoric stage of development (Johnson, 1997; Hann, 1988). Specifically, this interpretation of Indians' past assumes they lacked sustained commitments to teaching and learning, and that it was

Euro-Americans who introduced them to education. Recent research in multiple fields and revisited old documents, animated by interrogations of Indians' oral traditions, reveal the outsider-driven historical narratives of Native Americans to be shallow and, more to the point, incorrect. But to reconstruct more accurate and useable pasts, historians, including historians of education, need to frame a largely unprecedented scholarly discourse released from binary limitations. There is little evidence of sustained colloquia among Euro-Americans and Indians over the centuries that advanced cross-cultural knowledge (compare, e.g., Shannon, 2000; Odora Hoppers, 2002; Ostler, 2004). The normative meeting format has been a negotiation bounded by the long-established either/or choices of Euro-American centrism, with whites often bringing hidden agendas to the table. The approach has offered historians little more than a scholarly cul-de-sac that failed to lead to explanations of where and through what processes Indians learned and how so many managed to survive against horrific odds.

The Apalachees and Cultural Evolution

In 1987 a construction crew digging near the state capital in Tallahassee, Florida, happened upon deeply buried debris that looked strange. State archaeologist Calvin Jones inspected the site with growing curiosity. The artifacts being unearthed were indeed old—very old. Within months a plausible hypothesis took shape. Here were the remains of the first winter encampment of the Hernando de Soto expedition that landed near Tampa Bay in May 1539 (Miller, 1998). The educated guess proved correct, opening prospects of archaeological research on the only de Soto settlement discovered to date. Relying on 400-year-old reports on the expedition, archaeologists, anthropologists, and historians knew the Spaniards spent the first five months of their trek among the Apalachees. Now they could probe the site. Potentially this was an important advance. They had sketched a meandering de Soto trail across what are now ten southeastern states, but the exact itinerary has remained unknown.

Earlier Spanish attempts to establish outposts in the region had failed miserably. A record of one of these efforts provided details on an expedition also launched in *La Florida* by Pánifilo de Narváez in 1528 (Reséndez, 2007). Cabeza de Vaca, a survivor of the misadventure, penned a vivid account of the remnants' decade-long struggle against the uncharted waters of the Gulf of Mexico, harsh desert climates, and hostile, if curious, inhabitants. The small contingent ended up roaming through northern Mexico, south Texas, and territories to the west, but their disappointed quest for wealth and fame began with the Apalachees in northwestern Florida. The de Soto expedition failed too, but only after a four-year exploration by some 600 participants initially. It was the first recorded sustained encounter between Europeans and the indigenous peoples of what became the United States.

Scholars have been fascinated with de Soto as a dashing, ambitious, heroic figure, although they knew from the reports on his expedition that he left behind a wide swath of slaughter and enslavement as his entourage moved west (Ewen and Hann, 1998; Lamar, 1997; Avellaneda, 1997). Pathologists now suspect that the pandemics that raced through Indian societies in the sixteenth and seventeenth centuries can be traced to this Spanish invasion, the microbes hosted by its human participants, and the livestock it introduced to North America, notably domesticated pigs (Fitzhugh, 1985; Deagan, 1985; Mann, 2005). In the cases of the de Soto and Cabeza de Vaca documents, subsequent research has corrected and elaborated upon the original reports and narratives, only portions of which can be confidently accepted as the testimony of witnesses. Galloway (1997b) provides detailed readings of these old materials (also see Reséndez, 2007). Fully reliable analyses may never be possible. As now understood, both episodes have come to light through a convoluted chronology that moves between old written sources and contemporary research findings. A past gains form and substance aided and abetted by much more recent perspectives and technologies. This can be a troubling prospect for the history of American Indians, whose forebears in precontact eras tended not to leave written documents for posterity. The de Soto and Cabeza de Vaca reports provided glimpses of otherwise clouded cultural and social phenomena, but the revelations are cast from outsiders' angles. Alone, they cannot lead to unfettered accounts, and modern science can correct only some of the biases (Mann, 2005).

Profoundly affected by both the Narváez and de Soto expeditions, the Apalachees provide a tantalizing place to test a reconsideration. They were, after all, the initial targets of the two invasions and the Spanish hunger for gold that inspired them. Furthermore, researchers interested in precontact Indian cultures can now compare two kinds of sources, more or less contemporary documents describing the Narváez and de Soto encounters and ongoing investigations by a variety of scientists and social scientists, launched centuries later. With the Apalachees, whose post-contact history has been screened through European filters, primarily those of Roman Catholic missionaries, the methodological checks and balances of cross disciplinary approaches are invaluable. The veils obscuring their glory days can be pulled back partly, permitting at least plausible chronologies and explanations.

Cabeza de Vaca and the de Soto reporters described the Apalachees in similar terms (Reséndez, 2007; Steigman, 2005). The Spaniards beheld a handsome, prosperous, fierce people, whose central city, Anhaica Apalache, they expected to be another El Dorado, a place of great wealth and power. So reported the natives they encountered after landing at Tampa Bay. Shown shiny metals that looked like gold and pointed toward a fabled place where the treasure could be found, along with ample stores of food and supplies, Narváez and de Soto headed north, each in his own time. Their scribes pictured Apalachee lands as planned and well managed, with, in twenty-first century terminology, a major city, suburbs, exurbs, surrounding farmlands, and nearby forests with well-marked paths and little underbrush. A settled people, perhaps for as many as 500 years in northwest Florida, although the outsiders had no way of knowing this, they were organized centrally and hierarchically as a chiefdom, or so it appeared to the invaders (and early archaeologists). They had perfected the science of maize agriculture, grew other crops as well, and produced finely crafted pottery, baskets, tools, and weapons. Grievously disappointed, the Spaniards found no gold, misled probably by the Apalachees' talent for hammering copper into decorative and ceremonial objects. From repeated battles with the Creeks and other hostile neighbors, the Apalachees also acquired and refined military skills and a determination to protect their lands. They were unerring archers, daring and dogged warriors, and highly effective strategists, as both groups of Spaniards learned independently with grudging respect.

Archaeologists continue to confirm, correct, and add to the accumulated information. The Apalachees were probably direct descendents of the mound-building peoples who had occupied mid-continent territories as early as 1800 B.C. (Ewen and Hann, 1998; Hann, 1988). A diverse succession of civilizations, in later centuries they left traces of complex social organizations, replete with evidence of urban development, agricultural advancement, artistic achievement, and social stratification. They traded widely among themselves and other indigenous societies, perhaps as far south as Mesoamerica, a region with its own mound builders. Over the centuries, amazed inhabitants and travelers stumbled upon evidence of their existence almost by accident. Rumors spread among Indians and white newcomers. Archaeologists followed, quickly learning to identify the distinctive, durable footprints left in the form of earthen mounds, by the latest counts, thousands of them scattered across the country's center. Constructed by human labor, some were relatively small, but others reached heights of seventy feet and extended lengths, executing intricate patterns of animals, snakes, birds, or arcane religious symbols. The mounds served as ceremonial sites, burial grounds, or foundations for residences of the elite; in some instances they served multiple purposes. Before the end of the first millennium A.D., mound-building civilizations were scattered across the Mississippi Valley. Archaeologists surmise that they were Apalachee forebears (Hann, 1988; Johnson, 1997; Mann, 2005). This is more than a wild guess, given similar centralized polities, agricultural and trade practices, tool technology, and artistic creations. To a limited extent, the Apalachees also built mounds, but they were less imposing pyramidal structures used only for religious ceremonies. Located outside the major city, Apalachee mounds failed to catch the Spaniards' attention, possibly because by the early sixteenth century they had been abandoned.

The outsiders did notice, beyond the missing gold, the Apalachees' agricultural and military skills and their regional influence. The latter may explain why European map makers in the mid-seventeenth century attached their names to the highly visible and orienting geographical landmark that ranged northward from the southeast, the Appalachian Mountains (Hann, 1988). However, there is little documentary evidence that the invaders pondered the Apalachee culture. They relied on maize as a hedge against starvation and appreciated its benefits as a storable and transportable produce, but expressed little curiosity about the scientific knowledge needed to grow it. Respect came centuries later as biologists, botanists, archaeologists, and geneticists reconstructed the painstaking process of seed selection and cross fertilization that led to the invention of maize in Mesoamerica thousands of years before Columbus. Unable to reseed itself, maize was an odd crop, relying on human intervention at several stages. The achievement regarding maize cultivation required at the outset "determined, aggressive, knowledgeable plant breeders," qualifying, as one geneticist wrote in 2003, as "arguably man's first, and perhaps his greatest, feat of genetic engineering" (Mann, 2005, 196). This "bold act of conscious biological manipulation" sired offspring (Mann, 2005, 191–201). In relatively short order, maize agriculture moved north and south from Mexico, in each case necessitating adaptations to local climates and soils. The Apalachees had mastered the science long before the Spanish encountered them. Perhaps they brought it to northwestern Florida.

The Spaniards paid more attention to the Apalachees' military skill, courage, and ferocity. On several occasions, despite superior guns, metal armor, and horses, the Spaniards reportedly fell victim to guerrilla tactics and outright trickery. Yet they seemed unmindful of the learning and practice needed to execute battle-ready acumen. Researchers have now found evidence of the lifelong rehearsals for tactical war conducted among the Apalachees and the ever improving tool technology needed to manufacture effective weapons (Steigman, 2005). With native enemies nearby, they developed over the years what can be labeled a military culture. Competitive, sometimes fatal, intramural sports became part of it. The games were incorporated within rites of passage that tested adolescent males for bravery and determination against accomplished adults. The latter were expected to win. This is only one example of the grounds for speculating about the ways and means of cultural formation—education—among the Apalachees. Like their achievements within the military, none of their other achievements in art, agriculture, and trade can be explained as ad hoc occurrences. They imply cross generational and institutionalized commitments to intentional teaching and learning.

Conjecture about contacts between the Apalachees and Mesoamerica, including possibly the Triple Alliance (popularly known as the Aztec empire), point to additional fronts of cultural evolution during the long precontact era (Shaffer, 1992). If Apalachee traders traveled to the Mexican region, they very likely did so by sea, meaning they could design and build vessels capable of withstanding the treacherous Gulf streams, a notable accomplishment even if they sailed close to shore. They also needed experienced, confident navigators. Cabeza de Vaca and the other beleaguered remnants of the Narváez expedition, with Apalachees in hot pursuit, prayed to be blessed with just these seasoned skills, and God's good grace, as they hastily built rafts for their escape across the Gulf of Mexico (Reséndez, 2007). Only a few survived the voyage, landing amid natives in the south Texas region, who enslaved them. If by contrast the Apalachees had ultimately succeeded, among the bartered treasures they brought home may have been cobs of maize and the science of propagating it. There is fragmentary physical evidence to support such guesses. The Apalachees were indeed aggressive and adventurous traders. The Spanish testified to their skill and endurance as swimmers. It is also known that they routinely fished in the Gulf waters, but they could have amassed their agricultural expertise through sources closer to home or their mound-building ancestors.

Finally, there is the intriguing possibility that Apalachee traders witnessed or heard about the Triple Alliance's compulsory schools (Leon-Portilla, 1963). Flourishing during the fourteenth and fifteenth centuries and until the Spanish conquest in 1521, the institutions qualified as one of the world's earliest public education systems. All male children, whatever their status, attended one of two types of school. Each had its own curriculum, one leading to military and vocational service and the other to religious, political, or intellectual leadership. Although rank and prestige attached more directly to the latter, high born and commoner sons mingled at both institutions. The system shared general goals to inculcate the values portrayed in the empire's foundational and classical literature, including philosophy, myth, and poetry, and to promote patriotism and obedience to authority. The second purpose especially would have appealed to precontact Apalachee chiefs, who saw their people beset by envious marauders. There is no documentary or up-to-date archaeological evidence to suggest they established schools of either sort.

For unknown reasons, the centralized organization of the Apalachees began to fracture into more tribal configurations before the European encounters. Perhaps unrelenting war took its toll, or internal conflicts arose over the chief's authority and the distribution of status. Postcontact, an unexpected foe arrived hard on the heels of Spanish invaders. Disease ravaged Apalachee settlements, leaving them more vulnerable to their traditional Creek adversaries. Weakened and dispirited, some members of the community welcomed Franciscan friars in 1608, thus beginning the more traditionally documented missionary period (Milanich, 1994; Hann, 1988). The renewed Spanish connection, however, angered enough Apalachees to cause more fragmentation. Some migrated east to tribes in the Carolinas or were taken there by Creeks as captured slaves, others moved west. Traces of their presence have been found along the Atlantic seaboard and in Alabama and Louisiana, where by the early nineteenth century, probably through intermarriage, the distinctive Apalachees lost their tribal identity, and subsequent generations of Americans lost access to their oral tradition (Hann, 1988).

Coda: History as Presence

Resourceful and inventive, the Arikaras moved earlier and in a different direction. Part of the powerful, hegemonic Caddoan-language group that was centered across what is now Mississippi, Louisiana, Arkansas, and Texas, they began migrating upland perhaps as early as 1200 B.C. or as late as the fifteenth century (Ostler, 2004; Smith, 1996; Parks, 1991; Meyer, 1977). Historically related to the Pawnees, who also traveled north, the Arikaras may have started from as far south as the Mexico-Texas border, eventually settling in lower North Dakota along the Missouri River. Continuing into the present, they have honored the river as an irreplaceable part of their ceremonies. Their oral traditions and tribal stories do not reveal why or precisely when they relocated, although the people themselves cite divine guidance. Almost certainly they had departed before the de Soto entourage moved through the southeast region in the early 1540s. The expedition's reporters described encounters with the Caddos and some of their affiliated tribes, but they made no mention of the Arikaras. Not a people who could be easily overlooked, they lived in earthlodges clustered in permanent settlements devoted to farming and trade, practiced achievements that they took to North Dakota. On the way, before Euro-Americans arrived on the scene, they reintroduced maize agriculture to the Great Plains and upper Midwest regions, including several varieties of what is now known as corn. (Late mound builders engaged in maize farming, but their societies had long since declined and scattered.) Their initial journey stopped temporarily when they found advantageous sites along the Missouri for farmlands and a trading post. At the time, both were uncommon among Plains Indians and attracted attention. The latter became a center of intertribal trade among the Arapahos, Cheyennes, Commanches, Dakotas, Kiowas, and Plains Apaches, a place where the nations gathered, with the Arikaras functioning in the crucial role of peacekeeping "middlemen." Marking the beginning of the post-contact era in the Plains, white hunters later frequented the center as well. Arts, crafts, animal skins, and even traditional native dances were among the items bought, sold, or bartered. They continued this now traditional dual commitment to farming and commerce in subsequent migrations up the Missouri. The journals of the Lewis and Clark expedition described meetings with the Arikaras early in the nineteenth century (Parks, 1991; Ostler, 2004). Disaster struck in the 1780s and again in the 1830s when small pox epidemics spread in their settlements, a likely result of Euro-American contacts. Their numbers greatly reduced, the Arikaras nonetheless survived, as one traditional song put it, to "remember them, the ways of the old ones who were: The good ways that were ours" (Parks, 1991; see also Nichols, 2003; Fowler, 1996).

In 1991 Douglas Parks published a four-volume collection of their traditional narratives in the Arikara language, accompanied by English translations. Recounted by men and women with deep tribal roots, the stories tell of symbolic animals and mythic figures, episodes from the distant past, and memorable events of more recent vintage. Some tales rehearse tribal wisdom and values, often grounded in affirmations of spiritual gifts bestowed on the people by land, soil, corn, and the mysterious Missouri River. Others depict with subtle humor the Arikara practice of trying to mediate conflicts, whether among animals or humans, or across both worlds, as occasions required. Celebrated too are their entrepreneurial habits. They have long known how to drive hard bargains, intending to be fair, unless provoked by those with a different commitment. The narratives are meant to explain and to reawaken communal memory.

One of them reports "How Summer Came to the North Country":

Long, long ago when we people were not yet living on this earth, when the ways on this earth were holy, there was no summer here in this country. It was always cold then, always winter (Parks, 1991, 129).

Raven, Coyote, and Scalped Man, familiar characters in Arikara lore, complained, "everything sure is difficult for us when it is winter like this all the time, never getting warm, just winter, winter." They devised a plan to travel south to steal the Sun's son, which they did, racing back as far north as their strength permitted: "Now, there, is where the boundary of winter is, here where we are living today" (131). The story was passed along by community member Alfred Morsette. As Parks observes, it calls to mind the Promethean myth explaining the origins of daylight and fire on earth. It can also be heard (and now read) as an account that looks back on the Arikaras' historic migration from the southeast, and how they got there in the first place. The narrative portrays a tribal understanding of origins as processes, instrumental, to be sure, yet not just long ago events but seeds of an ongoing story. Perhaps they were returning to the "North Country," not discovering it. Or the meaning may be more literal. Arikaras could emerge from the underground, the cold, during the "holy" period when their traditions and institutions were still forming, because summer came north to melt the ice. They followed summer south, but over time the homeland drew them back to the place where the Arikaras began. In the several possible interpretations, the view of history remains directed toward the present, reconstructing explanations addressed to the people where they are "living today." Their journeys north brought "warmth" to other peoples along the way in the form of clustered learnings to be shared, maize, commerce, and the anchoring cultural habits and institutions that permanent settlements encouraged. It is a fascinating and empowering use of history, twisting time from the present to the past and back again.

These selected examples of precontact American Indian experiences point to processes of social organization and communal understanding that began in the undocumentable past. To learn about it, and from it, historians orchestrate an inventive mix of sources in the social sciences and the sciences, much of which require newly available technological tools. In addition, they expand traditional definitions of reliability and validity by weighing the utility of human memory in historical reconstructions. And here's the rub. Memory is personal, slanted, sometimes malignant, but always incomplete and malleable, not the usual stuff of hard fact, yet those deficits also qualify as benefits when triangulated with other sources. Memory can save historians from producing dull, unlived accounts that reflect a different order of bias and incompleteness. For one thing, it can take us inside experience, a particularly necessary excursion for education historians, who need to identify the phenomena of teaching and learning not only from the menus of intentions pronounced by self-named educational institutions but in whatever arenas they may occur. History itself, including education history, becomes processes of growing experiences from necessarily reexamined beginnings to the present. It enlarges known but inevitably unfinished worlds. Communal memory, rehearsed, exercised, and revised over years of practice, represents a uniquely muscular variant of the genre. Within it, encoded myths and other narratives open doors to inquiries inaccessible and even unimagined on trips along familiar paper trails. Far richer and more provocative than linear reconstructions, the resulting chronologies of American Indian history transform the past into a moving target, an expanding, inviting, educative presence.

Taken together, recent research, old documents, and oral traditions shed light on previously cloaked episodes of cultural formation among American Indians. European contact accelerated the processes and simultaneously very nearly destroyed them. Some Indian cultures evolved, as they had previously, surviving in altered, perhaps unanticipated forms. Others devolved or vanished. In all cases the multidisciplinary and multigenerational sources, and the perspectives they cast, support a working hypothesis: These were educated peoples.

CHAPTER 7

Researching Hispanic Undergraduates: Conceptual and Methodological Unease

Amaury Nora

Research on students' adjustments to and persistence in college has focused on identifying specific factors and the interplay among those variables based on theoretical frameworks. These conceptualizations, however, have not gone without some degree of disparagement. At the center of this discussion is the argument that the underlying assumption among such research endeavors is the notion of a "one-model-fits-all." Further, it is argued that these models examine the behavior of a very homogenous group—White males—and are not appropriate for other racial/ethnic groups. Along this same line of reasoning is the notion that findings from studies where students with very unique (and institution-specific) characteristics are used cannot possibly provide a "defensible" model. The question of generalizability has also led to the conventional view that the only way to determine what factors affect minority degree attainment and their adjustment to college is by examining large national databases. However, most national databases are dependent on proxy variables that do not sufficiently capture the important underpinnings of persistence models for minorities, and consequently they lack in rigor. Conceptually, this chapter will focus on the use of diverse conceptualizations to study minority student persistence through more racial/ethnic- and institution-specific datasets.

In addition to these conceptual problems that are associated with studying Hispanic students, there are methodological issues that are equally important to consider and address. Because we find an extreme degree of diversity among college students, the line of reasoning is that it is not possible to validly study this group of students through the use of the structured survey instrument alone. This chapter will scrutinize both issues by focusing on the expanding diversity of undergraduate students on our college campuses and its implications for research. Moreover, complex student-related issues make it difficult to select appropriate research and assessment methods (as well as theoretical models) by those who study college students. Finally, the intent of this chapter is to examine selected alternative research strategies that have been used to study the experiences of diverse student bodies relative to complex issues, particularly recent studies that have been conducted on the persistence of Hispanic college students.

The Presence of Hispanics[1] in Higher Education

The number of Latinos is increasing at a fast pace in colleges, and yet this group continues to trail other minorities in education (Fry, 2004), with only 10 percent of the Hispanics between the ages of twenty-five and twenty-nine having earned an undergraduate or graduate degree as compared to 34 percent of their White counterparts and 18 percent of Blacks (Llagas and Snyder, 2003). The largest subgroup of Latinos, Mexican American students, drop out of high school at a 50 percent rate when using 9th grade enrollment as the baseline year. Among those Latino students who make it through the K-12 system, not all who graduate manage to enroll in college. Only 35 to 40 percent of Hispanic high school graduates enroll in higher education (Arbona and Nora, 2005; Nora, 2005). Among the general population, roughly 22 percent of the eighteen to twenty-four-year-old Hispanic students in the United States attend college. The vast majority of Hispanic students that are eligible to attend college are enrolled in two-year institutions (Fry, 2004). While it is estimated that 36 percent of Hispanic students enroll in college following their high school graduation (an increase from the 27 percent in 1985; Llagas and Snyder, 2003), 48 to 55 percent of White students graduating from high school go on to enroll in two-year or four-year institutions.

Even when Latino students express a desire to enroll in college, issues related to college choice and access create barriers and impediments for Hispanic students who do enroll. While the issue of access into higher education for Hispanic students has been declared a national priority at all postsecondary levels by the Clinton and Bush administrations (Fry, 2004), regrettably, large gaps in postsecondary participation exist and much more research is needed regarding the unique needs of Latino students in higher education to make a difference in the current disparity (Hurtado and Ponjuan, 2005). Affecting the entry for Hispanic students requires an awareness of existing issues related to true access, as well as an awareness of the on-campus and off-campus experiences that impact the persistence of this group. An understanding of the multifaceted components of Latino students' college experiences is recognized as vitally important (Castellanos and Jones, 2004).

Studying Latino College Students: *Theoretical Unease*

Early in the study of Latino undergraduates, scholars (e.g., Murgia, Padilla, and Pavel, 1991) focused their attention on diverse student background characteristics and membership in student enclaves. These student structures are important because they afford students a way of coping with the complicatedness of college life through social support and strategies for living. Moreover, enclaves influence the manner in which students interpret events and problems, providing them with attitudes and values, on the basis of which they develop dependable patterns of responding to and fitting in campus events. The distinctiveness of these student groups on a campus is the product of institutional and student characteristics (Kuh and Whitt, 1988). Students' precollege characteristics (e.g., values and attitudes) and prior acquaintance with one another, as well as their post-matriculation characteristics (living arrangements, class, and close relationship with an organization), influence the development of subcultural groups, as do institutional culture, significant others within the institution, administrative structures, and the size and complexity of an institution.

One major criticism of categorizing students, specifically students of color, into such typologies is that such subcultures are largely heuristic or analytic tools used to exemplify the notion of student subcultures in higher education and are not descriptions of specific racial/ethnic groups on a particular campus (e.g., Hispanic college students). Another criticism is that categories of these typologies often fail to meet generally accepted criteria for subcultures, such as persistent interaction, processes of socialization, mechanisms for social control, and norms that differ from the parent (overall institutional) culture (see Bolton and Kammayer, 1972; Horowitz, 1987; Van Maanen and Barley, 1985; Warren, 1968).

However, the existence of these enclaves on different campuses and the membership of Hispanic students in those subcultures produce fundamental differences among Hispanic students in their post-matriculation orientations toward a college education and begin to suggest the diversity of campus peer groups to which Hispanic students can become attached. The influence of membership in these subcultures among Hispanic students is pervasive. It would be appropriate and conceptually wise for researchers of Latino students to take these factors fully into consideration.

New Theoretical Perspectives:
Casting a Conceptually/Culturally Sensitive Wider Net

More recently, quantitative survey tools and qualitative analyses surrounding students in higher education have been concentrated on how college students engage themselves in the classroom and on campus. The research has led to overviews of the major constructs/factors that have been found to impact the adjustment, academic performance, persistence, and graduation of undergraduates in higher education, specifically of Latino students (Nora and Crisp, forthcoming; Nora, Barlow, and Crisp, 2005). The findings are helpful in not only identifying enrollment gaps and persistence rates but also conceptual models that point at the underlying structural patterns among factors that have an impact on different student outcomes. More importantly, these more recent models have been instrumental in calling attention to what still remains to be learned regarding Hispanic higher education students.

Although there has been a growing number of Hispanic scholars who have contributed significantly to this body of literature and have opened lines of research that encompass a more meaningful understanding of postsecondary Latino(a) students (e.g., Castellanos and Gloria, 2003, 2006; Gonzalez et al., 2006; Hurtado, 1997, 2006; Olivas, 1995, 2005; Nora, 2001, 2002, 2004; Padilla et al., 1997; Rendon, 2000; Venezia, Kirst and Antonio, 2003), there are still specific areas in the literature that have not been fully, theoretically, and empirically investigated.

In a recent gathering of Hispanic scholars and administrators, three key issues were identified as unexplored areas regarding Latino undergraduate students (Nora, forthcoming). The first topic is the need to examine current definitions of success that are used as the standard in providing measures of goal attainment, accomplishment, or desired outcomes in research studies on Hispanic students. These designations of success fall short in representing appropriate measures of the construct under investigation—success in higher education—and their soundness in capturing the same conceptual meaning among different racial/cultural student populations. Researchers have emphatically noted that definitions of success need to be redefined so that they are more appropriate in their application to different student groups.

A second major area for discourse is the inclusion of noncognitive measures in databases rather than simply focusing on cognitive outcomes such as grades or graduation rates. No one denies the importance of academic grades, enrollment and retention rates, and graduation counts. On the other hand, these cognitive-related outcomes do not occur in isolation of student attitudes, their values and perceptions, and their on-campus and off-campus academic and social behavior. It is naive to think that the interactions among and between students and faculty do not have significant bearings on all student outcomes. Take, for example, the racist theme parties that exist on college campuses. How do they not have an impact on student learning, aspirations, motivations, and integration? Aren't a student's sense of belonging on a campus or the observation that faculty not only encourage but fully involve students within their classrooms factors that have an influence on student departure? In conjunction with the call to focus more on longitudinal models rather than cross-sectional views, there is an equal urgency for behavioral and noncognitive data to be measured and included in large national, longitudinal data-gathering efforts.

Another consideration in the study of Hispanic college students is the need to develop fully comprehensive, theoretically driven, and culturally sensitive databases for every sector of higher education. Much of the research on minority students has depended primarily on the use of extant databases, consisting of measures that are based on ethnocentric definitions and conceptualizations of important constructs. There are those investigators, such as Rendon et al. (2000) and Tierney (1992), who have disparaged the use of current models to study racial/ethnic student groups. Their contentions focus on the inaptness of variables to depict the complex differences, culturally and ethnically, of Latinos, African Americans, and Asian American students. While these judgments may be valid, the real condemnation is not in the use of constructs (or theoretical models) across different groups but the manner in which the items that measure these factors are constructed so that they more accurately, culturally and differentially, capture student experiences and perceptions of the group under study. For instance, current models of student persistence emphasize the importance of academic and social integration on campus. The argument should not focus on whether those constructs are functional for all groups but, rather, how the measurement of those constructs can capture the cultural and ethnic differences of the groups. How do Latinos, Blacks, and Whites differ in the way they academically and socially integrate themselves on campus? The degree or magnitude to which these different groups integrate can be measured, and its impact on student outcomes tested, only after researchers make sure that those variables are representative of the constructs under consideration for each group.

Conceptually, research in education and in the fields of psychology, anthropology, and sociology are altering the way in which we conceptualize different observable facts and the manner in which we select empirical tools to guide our observations and investigations (Hurtado, 1997; Rosaldo, 1989). Hurtado's (1997) research on feminist issues calls for a multidisciplinary and multi-method approach that is nonlinear and where one prevailing group is not privileged over other groups. It also invites critique and subsequent analysis. Based on these two circumstances, research on Latino students cannot remain entrenched in traditional and inappropriate theoretical frameworks. More contemporary models and perspectives and theory refinements are needed, that is, conceptualizations that consider the central theoretical issues associated with the experiences of Latino students in higher education.

Rendon, Jalomo, and Nora (2000) point out that innovative and theoretically sound notions regarding Hispanic student persistence had previously failed to surface in the literature. The prevailing view among many researchers and practitioners is that the retention of Latino students is similar or identical to that of majority students. This perspective leaves us with almost universally entrenched views that existing models of dropout behavior (including the assumptions on which they are based) are complete, appropriate, and valid for Hispanic students, regardless of economic and social backgrounds. As previously noted, the research by Hurtado (1997) contends that linear models based on an assimilation/acculturation framework leave many questions unanswered, mainly with regard to multiple group identifications and the way that both minority and majority groups change when they come into contact with each other. The adherence to a "one-model-fits-all" assimilation/acculturation framework simply maintains that students assimilate to what are typically White-dominant norms on campus. It is hoped that researchers and research will consider the inclusion of new, and more appropriate, perspectives and frameworks in studying the success of Latino college students.

Researching a Diverse Student Body: *Methodological Unease*

There is an argument that students from different ethnic backgrounds are expected to be culturally different. What is more, students who share the same ethnic culture are also likely to differ subculturally because of gender, age, socioeconomic or educational status differences, or their campus peer group membership. Kuh and Whitt (1988) identify two levels of operation for students. One is that of the general culture, in which common "collective, mutually shaping patterns of norms, values, practices, beliefs, and assumptions" are directing their behavior and providing them with "a frame of reference within which to interpret the meaning of events and actions on and off campus" (2). The other level is the subcultural level, where the shaping factors are quite different, particularly for students of color.

As previously noted, researchers of college students must contend not only with a diverse population of students but also with a diverse and complex set of conceptual issues. As a result, to meaningfully explain why Hispanic students exhibit various levels of retention and adjustment, or to satisfactorily assess their academic achievement, it is necessary to consider more than one or two potentially influencing factors at a time. Prior reviews of the research literature on student retention (Attinasi and Nora, 1987; Pascarella and Terenzini, 1995; St. John et al., 2001) revealed how the failure to do so for years slowed progress in the study of student attrition. Isolating on a single variable, be it financial aid, college preparatory courses, educational aspirations, and so on, can only lead to the misspecification of conceptual frameworks and to the exclusion of nonexperimental research designs and sophisticated multivariate data analyses. The use of longitudinal research designs to gather data and quantitatively test hypotheses and models through such techniques as structural modeling, hierarchical linear modeling, and logistic regression analysis take into account the inclusion of a multiple set of variables at a single time.

As more attention is paid to minorities in higher education, it becomes clear that many areas of interest to researchers and evaluators of Hispanic students have neither come to light nor have been fully examined. Some areas are only just beginning to be investigated in terms of theoretical perspectives or conceptual frameworks. These new theoretical structures should serve as the underpinnings of explanatory models, which are multifaceted in nature, of such phenomena as Hispanic college student retention, transfer, and academic performance.

Moreover, identifying and clarifying the processes leading to various outcomes of Latino college attendance should not be limited to findings or data based on a quantitative paradigm. Relying on qualitative (naturalistic) research, such as ethnographic studies, will result in culturally- and racially sensitive frameworks, which then can be further subjected to quantitative investigations with larger numbers of Hispanic (or other) students. Conversely, testing specific quantitative models may generate new hypotheses regarding the interrelationships among factors in existing conceptual frameworks that can be examined much more deeply through qualitative research. Even with the current emphasis on scientific evidence and methods, naturalistic research is capable of producing findings that could enhance our understanding of the course of actions underlying cause-and-effect relationships established through experimental and quasi-experimental research. The main point to be made is that researching Latino student outcomes and evaluating student progress should be conducted with a variety of research methods.

Methodological Tools

With considerations of cultural and subcultural diversity as well as with individual subculture, even cultural uniqueness, methodological issues such as the use of planned survey questionnaires that are constructed largely from the perspective of the researcher become problematic. Quantitative researchers often assume, even if it is implicitly, that they are the experts and know what is important to ask regarding college student experiences. Such convictions often result in denying the possibility of alternative findings. The use of survey tools alone cannot lead us to an understanding of the experiences of Latino students in institutions of higher education.

Researchers must be open to the acceptance and use of qualitative tools as, for example, ethnography, which is a methodology appropriate for cross-cultural research. Most significantly, this methodology provides a perspective from within as to what is happening in the group's natural setting. This emic perspective is often revealed through the use of research methods such as participant-observation and in-depth interviewing. These methods allow the researcher to understand an experience (e.g., what it is like to be a Latino student in a largely White institution) from the point of view of the student and without the interference of prior researcher conceptions. Only after revealing the student's perspective through qualitative data collection is the researcher in a position to clarify his or her observations from an outsider's, or etic, perspective.

One of the most significant contributions of open-ended research techniques (i.e., participant-observation and in-depth interviewing) is that they permit the researcher to examine culturally different people in an open-minded way. The use of qualitative methods to study the Latino college experience is key to openness to frames of references that are different from those routinely used by researchers of college students, and makes it particularly suitable for investigating Hispanics and African Americans, which are culturally and subculturally diverse student bodies.

Different Methodological Approaches to the Study of the Retention of Hispanic Students

The study of student retention is one that has been scrutinized by both naturalistic and quantitative (survey) approaches. Nora and associates (1996, 2000, 2008), Gonzalez et al. (2003) and Rendon (1994) have all conducted studies on the persistence of undergraduate Latino students. The use of various methodologies (i.e., multivariate analysis, structural modeling, and naturalistic research) in these investigations have helped to enhance our understanding of Hispanic student retention, a phenomenon that is multifaceted not only because of the complicated nature of the persistence construct itself but also because of the additional aspect of the cultural uniqueness of Hispanics. Both paradigms have made unique contributions through in-depth interviewing of persisters and nonpersisters and causal model testing with survey data and the potential contribution of their integration.

Descriptive studies, while informative, do not propose potential causes of Hispanics' underrepresentation and underachievement in institutions of higher education, but their profiles make clear the call to move beyond statistical description of a phenomenon to an understanding of why the phenomenon exists and what can be done to improve it. They also do not address such questions as: Why are Hispanic students so underrepresented in undergraduate degree attainment in every area of higher education? Are high attrition rates among Latinos a matter of educational aspirations, precollegiate attributes, financial circumstances, institutional factors, or the interaction of all of these things? Are motivations affected by characteristics specific to college choice? What insights about the condition of Latino students can we arrive at by more direct and in-depth discourse with those who are affected? Effectively addressing these issues requires that we come to depend on both quantitative and naturalistic studies of Hispanic experiences in higher education for insight.

New Perspectives on Familiar Paradigms: Quantitative Voices and Stories

Despite opposing views, and the acknowledgment that I am a quantitative researcher, I truly believe that quantitative coefficients, parameter estimates, and numbers tell their own, and exciting, stories. The use of such statistical techniques as structural equation modeling, logistic regression, or hierarchical linear modeling to assess the appropriateness of hypothesized causal models should not simply establish an association between numbers. When there is a great deal of research and forethought to capture and represent the real meaning of variables within a quantitative model, the magnitude and direction of a path or standardized beta coefficient as it relates to the interaction of two factors uncovers a story as it is played out in real life.

For example, Nora and Cabrera (1996, 1999) tested a causal model that would examine differences among minorities and nonminorities as to what factors impacted their persistence in college. One of the variables in the studies attempted to capture the perceptions that students had regarding discriminatory behaviors and gestures on a college campus and in the classroom. As might have been expected, a negative relationship was found between perceptions of discriminatory acts and attitudes and a student's decision to remain in college or drop out. In other words, if students believed that discrimination was seen in the classroom or in different parts of the campus environment, the likelihood that the student would remain enrolled was diminished. However, that was not the full story that was told by the quantitative findings. Another variable found in the hypothesized model was a measure of the degree of support and encouragement that students had received from significant others in their lives, mainly parents.

The statistical coefficients also revealed that the negative impact of perceived discrimination on a college campus was offset or negated by the support and encouragement from significant others. These findings are represented by gammas, betas, direct and indirect effects in a quantitative table, but there is a story to be told by those figures. Even though students may face the sick reality that discrimination and bigotry exists at the institution in which they are enrolled, so much so that it is found not only in cafeterias, lounges, and residential areas but also in the classroom, their social support systems or safety nets that are provided by parents (or significant others) will counter the inclination to give up on college or fleeing to a different campus. Words of encouragement and support keep students from dropping out.

Add to the story the inclusion of several more variables/factors/constructs (and their interrelations) and the stories that are told become even more intricate and exciting. Quantitative findings are not (and should not be) simply the degree of variance that two variables share in common. That is not how reality works. Rather, there are numerous factors, all linked together in different ways, that, when combined, unveil those student stories. The stories are there, found among the numbers, waiting to be told by the researcher.

How is this discussion related to researching Hispanic undergraduates? The incorporation of pertinent variables in theoretically sound causal models and the simultaneous testing of those variables quite often reveal startling findings among different student groups. For example, very early in the study of student retention, Nora (1987) substantiated the relationships hypothesized in a model of student persistence and concluded that the causal model was a plausible representation of influences on student retention specific to Hispanic students in two-year institutions. Although the model explained a good deal of the variance in Hispanic student attrition, several factors found to be significant in predicting student persistence for majority students in four-year residential institutions did not appear to have an impact on the retention of the Hispanic community college students that Nora studied.

The findings only partially supported the hypothesized relationship between measures of academic integration and retention that had been found in previous retention studies. However, there was more to the story being told. That same analysis further revealed that a relationship between measures of social integration and retention was not substantiated for Hispanics in two-year institutions, although social integration had been found to have the largest influence on withdrawal decisions for majority students. In the end, the researcher found that, for Hispanic students, measures of initial commitments (to the institution and educational goals) had significantly large impacts on Latino student retention.

One of the major criticisms of quantitative studies is that they do not incorporate student voices, and that the survey items and the constructs they represent are more indicative of the thoughts and biases that the researcher brings to the study. It is sad to say so, but this is more often the truth than not. And, while that may be true of variables in large national databases, forcing researchers to rely on proxies for important latent constructs in their research, it is not necessarily true when student surveys are used at the institutional level. Nora and Gracia (2000), Nora and Crisp (2008), Bensimon and Nora (2000) have all used qualitative data from interviews to guide the development of items for future data analyses. In all three cases, transcriptions from individual interviews with students or from focus groups have been used to identify not only the major themes that represent student voices but also the phrases and language used by students in the writing of the items for the survey tools. No preconceived themes or constructs were introduced in the qualitative data gathering or in the development of the survey instrument. The intent was to incorporate those thoughts, perceptions, and attitudes that were voiced by informants and subsequently tested on the larger student population. In this way, the dimensionalities of the constructs that made up the quantitative causal models were better represented conceptually and culturally, and made racially sensitive.

New Perspectives on Familiar Paradigms: Generating Grounded Perspectives through Qualitative Studies

Data collected through in-depth, open-ended interviewing are analyzed inductively, often to generate grounded concepts for interpreting the context within which students make decisions and display student outcomes. Most analyses often start with open-coding of the interview transcriptions, or coding contents in as many ways as possible. Coding categories may relate to ways of thinking about people and objects, processes (sequence of events, changes over time), activities, events, strategies, relationships, and social structure. Data is subsequently reduced as to the number of coding categories, and the analysis begins to take shape conceptually. Judged by the frequency of occurrence, its uniqueness and apparent connectedness to other categories, various categories are retained, merged, and/or discarded. It is also possible to link categories to one another through higher order categories. These higher order categories, or concepts, are subsumed under another category but without sacrificing their integrity. Initial coding categories become subcategories or properties of higher level categories, with eventual connections established between them.

For example, a naturalistic study on Hispanic students by Attinasi (1991) revealed two conceptual schemes: "getting-ready" and "getting-in." The getting-ready concept included acquiring early in life through parental communication an expectation of going to college, witnessing early in life the college-going of family members, having vicarious experiences of college through communications from high school teachers, and having direct experience of college through prematriculation visits to college campuses. Getting-in included post-matriculation experiences associated with the student's management of the new environment. Embedded within the concept of getting-in were two subconstructs, "getting-to-know" and "scaling-down." Getting-to-know involved acquiring familiarity with the physical, social, and academic spheres of the campus through contacts with veteran students and through information sharing with fellow first-year-in-college Latino students. Scaling-down represented the tendency of Hispanics to confine their activities to narrow portions of the three campus spheres (physical, social, academic) in effectively reducing the amount of new knowledge they have to acquire to successfully negotiate them. As persisters and nonpersisters differed in terms of both getting-ready and getting-in experiences, Attinasi concluded that these concepts were important for understanding Hispanic student retention.

New Perspectives: Combining Paradigms to Inform Each Other

Research on college students has typically involved either the confirmation of relationships between variables (through quantitative survey methods, for example) or in-depth examination of behavioral processes (through qualitative, naturalistic research methods). There are times, though, when it may be desirable to integrate the two approaches in order to address the issue of unexplained variance in quantitative models and enhance understanding of relationships established between factors in these models. When tested empirically, causal models of phenomena, such as college student retention and achievement, do not often account for a large amount of the variance in a behavior (e.g., persistence) of interest. It is possible that the large unexplained variance associated with most of these models could be due to either (a) the failure of the researcher to adequately capture the true nature of the factors specified in the conceptual framework from which the model is constructed or (b) some misspecification of factors in the conceptual framework itself.

Consider, for example, a theoretically driven quantitative model in which a variable, say the student's commitment to his or her institution, is hypothesized to directly affect a student's persistence in the institution. If after repeatedly testing the hypothesized model the researcher consistently finds no direct effect of the measure(s) of commitment on the measure(s) of persistence, he or she might be tempted to conclude that a direct relationship between the factors is misspecified in the conceptual framework. This conclusion is unjustifiable, however, as it is equally plausible that the failure to find a direct relationship between the factors is due to inappropriate operationalization of one or both of them in the causal model.

The need for concern is that the exclusion of relevant variables from a causal model most often leads to exaggerated or underestimated effects and a reduction in the variance accounted for by variables in the model (Pedhazur, 1982). Even when variables that are not associated with other variables in the model are excluded, it can have adverse effects because, by increasing error terms (unexplained variance), such exclusion can distort the relationships among variables already found in the model. As is most often the case, important cultural or culturally sensitive variables are excluded in studies of Latino students.

One way to counter the misspecification found in quantitative models of college student outcomes, such as persistence, is to begin to incorporate into these models findings from naturalistic research (Fry, Chantavanich and Chantavanich, 1981). Qualitative research has the potential for identifying factors that influence outcomes that will otherwise remain unidentified, for this reason unmeasured, and as a result subsumed under error terms in the data analysis. The inclusion of factors that belong in conceptual models, as well as appropriate measures of them in quantitative models, should not only reduce the risk of specification error but also increase our understanding of the processes underlying relationships established in these models.

The above logic should be undertaken in studies seeking to improve the explanatory capacity of quantitative models, particularly as they apply to minority students. These studies would combine theoretical frameworks with qualitative factors identified in naturalistic investigations. The result should be causal models that include new measures of components based on omitted concepts. New causal models would result in better overall fits of quantitative models in explaining student outcomes. New, perhaps better, operationalization of components of theoretical frameworks improve the measurement of those factors that would be subsequently tested with survey data. Naturalistic research also has the potential for providing in-depth understanding of relationships between factors that are well established through the testing of causal models with survey data. For example, although we might have established a statistically significant relationship between particular quantitative measures of, say, social integration and persistence through causal model testing, we might still wish to understand how the relationship plays itself out in the everyday life of individual students; after all, there is only so much of human behavior that can ultimately be captured in numbers. To achieve such understanding is the object of naturalistic research.

Conclusion: The Need for Better and Culturally Sensitive Research on Latinos

There is no denying that the increasing diversity among college students has added to the complexity of the issues that impact student outcomes, particularly for Latino and Black students. States such as

Texas have even set up special task forces (e.g., Texas Higher Education Task Force on Latino Students) to deal with the gap in college attendance that exists between Hispanic students and majority students, mainly created from the judicial decisions, such as the one in *Hopwood v. Texas*,[2] and ineffective attempts (Texas Ten Percent Plan[3]) to correct enrollment gaps primarily in flagship universities. To address the underrepresentation of Latinos in higher education, it will require research findings that speak directly to barriers and impediments faced by these students on different campuses.

Just recently, the Supreme Court depended on evidence establishing that diversity of students is beneficial to all students and used it as the sustaining argument for their ruling on *Grutter v. Bollinger*;[4] this evidence was produced by studies on Hispanic and Black undergraduate students. With so much of the decision riding on "scientific" evidence, it is not difficult to imagine the conceptual and methodological unease researchers face as they engage in their empirical investigations on such outcomes as persistence, academic achievement, and degree attainment for Latino students.

The cultural and subcultural diversity of Hispanic students calls for the use of diverse research methods that allow the researcher to be sensitive to diverse frames of reference, many of which may be quite different for different racial/ethnic student populations, and also differ from the investigator's own. A standardized questionnaire will not adequately capture the experiences and attitudes of minority students who have diverse racial or ethnic backgrounds or who are subculturally diverse due to differences in age, sex, special needs, financial status, or campus peer group affiliation. Rather than relying on a single approach, researchers should be able to draw upon the rich research methodology developed by anthropologists and others who have traditionally engaged in cross-cultural research and ground their understanding of what happens to Latino students in college in the group's own understanding of events. The upshot of efforts and policies aimed at closing the gaps for Hispanic students will largely depend on conceptually and methodologically sound research that is culturally sensitive to the diversity of these students. Such evidence will be required even more in the next court battle on affirmative action and minority students.

Notes

1. The terms Hispanic and Latino will be used interchangeably throughout the chapter. Both terms have been used in quantitative and qualitative studies in

higher education referencing Mexican Americans, Puerto Ricans, Cuban Americans, and those from Central and South America.

2. *Hopwood v. Texas* led to the end of preferential treatment with regard to the admission of underserved minority students to higher education institutions in Texas.

3. In response to the decision of the *Hopwood v. Texas* court case, the Texas Ten Percent Plan was drafted as a race neutral admissions policy. The statute ensured students graduating in the top 10 percent of their graduating class from any high school in the state of Texas that they could attend any higher education institution in the state, including the University of Texas at Austin and Texas A&M University, the state's flagship universities.

4. The ruling on affirmative action by the U.S. Supreme Court in *Grutter v. Bollinger* made it possible to use affirmative action policies in admitting minority students at the graduate level.

CHAPTER 8

Get Real: The Process of Validating Research across Racial Lines

Rachelle Winkle-Wagner

Introduction

Norman Denzin and Yvonna Lincoln (2003) introduce their edited volume, *The Landscape of Qualitative Inquiry* by recalling a famous picture of a White researcher studying a community in India as an example of cross-racial-research-gone-wrong. This picture also provided the cover to an important book on ethnography called *Writing Culture* (Clifford and Marcus, 1986). In the picture, the white anthropologist (or fill in the appropriate discipline) is sitting on a chair away from three dark-skinned people, looking in on a child, a woman hanging back in the shadows, and a man who is staring back at him while the researcher shields his White skin from the sun with a symbolically chosen white handkerchief (Denzin and Lincoln, 2003). He does not interact with the man, child, or woman whom he observes. Rather, he takes notes, the quintessential picture of the "researcher-as-expert," or the "lone ethnographer," and he will then be deemed the expert of this community or cultural group. It is as if the notes that he takes—White looking in on Brown (read "other")—will be able to seemingly uncover something that those within the community would never explore or could never understand for themselves. Implied by this picture is that the *separation* maintained by the researcher and the subjects of his gaze is the only way that this researcher could possibly be valid.

Scholars across disciplines have since reacted to this notion of researcher-as-expert, and of White-observing-Brown or White-observing-Other (Gunaratnam, 2003; Bridges, 2001; Merriam et al., 2001; Banks, 1998;

Zinn, 1979). In response, some assert that cross-racial research, with its historical baggage of exploiting and often misinterpreting the "other," should not be conducted at all, or at the very least should be challenged (Johnson-Bailey, 1999; Newsom et al., 2001). Yet, this also seems problematic. If White people cannot do research across race, will White people ever confront their own privilege, historical power, and the culpability for that? If White people cannot do research outside of their racial group, does this mean that the bias toward only including White participants in research findings will continue? Or, does this mean that scholars of color must only study the notion of "race" as it relates to their own racial groups? What does this mean for cross-racial understanding if everyone only studies those in their own racial group? Yet, if this research across racial lines is conducted, how can the work be validated given differences in backgrounds and cultural contexts?

While the picture referenced by Denzin and Lincoln (2003) produces a visceral reaction within me of distorted power in the research process, of cross-racial research gone horribly wrong, this was the way that I was initially taught how to conduct research. One should be "objective," not too close to participants. One should be the expert on a particular culture, phenomenon, social group, or issue with the implicit assumption that the "subjects" are somehow incapable of being the experts of their own lives and experiences. There is hardly a need to *talk* to the people because a researcher, as if in some infinite wisdom, can simply *observe* behaviors and generalize them, making claims as to the norms, values, and group processes of a cultural group. Research would be valid if the *researcher* said it is so—the participant could not possibly understand the validity of her own meaning making. Of course, Denzin and Lincoln (2003) point out the un-confronted paternalism, power, colonizing nature, and naivety of this notion of research, and I largely concur with their position. Yet, I am not convinced that this problematical notion of research, particularly for White scholars researching people of color, has been discussed enough. Particularly lacking is a consideration of the process of validation across racial lines.

In this chapter, I grapple with the issue of validation[1] in cross-racial qualitative research and in critical ethnographic work in particular. I reflect on my own challenges of validation as a White researcher conducting a study with African American college women (Winkle-Wagner, In Press). I first provide a brief description of the history of relevant qualitative research, which provides the backdrop to these current challenges and struggles related to research across racial lines. Then, I delve into issues related to research validation, before weaving together these two areas to understand the validation of work across racial lines.

Colonizing Research and the Ethics of Other-ing

Historically, qualitative research, initiated in the anthropological tradition, came of age through studies conducted by European/White researchers on colonized groups. Consequently, these studies were often a study of the colonization process itself rather than studies of cultural or social norms. Often researchers, by failing to interact in meaningful ways with the participants, actually reinforced or *colonized* a culture into the researcher's cultural norms, rather than understanding the culture from the participants' perspectives. Indeed, the history of the ethnographic tradition is steeped in colonialism, dating back to the seventeenth, eighteenth, and nineteenth centuries (Denzin and Lincoln, 2003).

Colonizing research, or qualitative research as it was historically conducted (and arguably the way it still *is* conducted in many cases), was intimately linked to the perspectives of the conquerors, those who annihilated indigenous cultures. Simply stated, "The colonial model located qualitative inquiry in racial and sexual discourses that privileged White patriarchy" (Denzin and Lincoln, 2003, 48). Thus, the written reports of these research efforts represented non-European groups (i.e., Africans, Asians, and Third World peoples) as "other" or "alien" or "abnormal" or somehow inferior (Geertz, 1988). This period of qualitative inquiry, often referred to as the period of the "lone ethnographer" (Rosaldo, 1989) or the "traditional period" (Denzin and Lincoln, 2003) maintained the researcher-as-expert, or the researcher-as-objective, encountering the people studied from a purportedly "value neutral" stance. Those being researched in this case were often treated as though they were objects. As such, the people being studied had little influence over the research process and were reported to have little influence or power over their own lives (for example, see Levi-Strauss' depiction of indigenous cultures as "savage," 1966).

Although more implicit, these roots of colonialism still influence contemporary qualitative research. More recent ethnographers and qualitative researchers have attempted to separate themselves from this tradition of colonizing and other-ing (Vidich and Lyman, 2003). There is a general, and indeed quite legitimate, distrust of cross-racial research. Often cross-racial work is a mirror to the history of colonizing, particularly regarding research conducted by White researchers on people of color.

Scholars employing action, critical, and feminist research perspectives responded to and distanced themselves from this colonial tradition of qualitative inquiry, attempting to democratize, humanize, and equalize the research process—ultimately attempting to respond to the other-ing that occurred in earlier forms of qualitative inquiry. These three traditions are complementary to one another rather than mutually exclusive. They all challenge the position of the researcher-as-expert while simultaneously attempting to position the research participants to be in control of their own lives and lived experiences. In the case of action research, there is an emphasis on praxis (i.e., social action) and actionable change as an outcome of the research process (Greenwod and Levin, 2003). Feminist (Olesen, 2003) and critical research (Kincheloe and McLaren, 2003) both problematize the White, patriarchal historical tradition, attempting to simultaneously uncover oppressive forces while also working to ameliorate these forces in and through the research process. However, there can also be action-oriented aspects of critical and feminist research.

In addition, as a response to the earlier other-ing and colonizing in the qualitative research tradition, there has been a shift regarding the role of the researcher. This shift challenges the value neutrality of researchers, particularly White researchers who study participants of color. Researchers grappling with the role of the researcher assert that claimed value neutrality may, often unwittingly, become a representation of the values of the mainstream White patriarchy, leading to other-ing and de-normalizing of participants (Fine et al., 2003; Carspecken, 1996). Thus, the role of the researcher has shifted in qualitative research from the lone observer to a more participatory role (Denzin and Lincoln, 2003), that of a participant observer, active participant, or advocate.

It is no wonder, given the biased history of cross-racial work, that there is so little discussion of validation across racial lines. If the performance of cross-racial research is still somewhat contested, the validation of this work might be even more contentious. Some may even consider the validation of cross-racial work impossible. Within this context of qualitative research, colonizing, other-ing, and later efforts to equalize and democratize, I reflect on my own methodological journey as it relates to research across the color line, particularly focusing on the challenges of validation within this process.

Approaches and Techniques for Validation in Qualitative Research

The issue of validation, particularly in qualitative work, is often contentious, nebulous, and criticized. This is perhaps one of the most important issues in research methodology because the validation of research in many ways, as the name suggests, underscores what is "valid," valued, recognized, or accepted as legitimate research. The objectivist epistemology[2] upon which quantitative work is based provides the dominant framework for what is often considered "valid" in research processes. As the chapters by Apple, Carspecken, Delandshere, and Dennis maintained at the beginning of this volume, the primacy and dominance of "scientific" or positivist[3] perspectives to research has influenced research across the board. Likewise, this dominance of positivism has affected views and approaches to validation in both quantitative and qualitative research.

Not only are there a variety of types of validity (Maxwell, 2002; 1992),[4] but there are also numerous validation techniques employed in qualitative work. Validation techniques in qualitative work often mirror the epistemological stances, or what some scholars consider the "lens" of researchers (Creswell and Miller, 2000, 125; Mays and Pope, 2000). Some scholars, primarily those using an objectivist epistemological stance, assert that qualitative research can and should be validated in the same way as quantitative work, the terms just need to be operationalized differently (Mays and Pope, 2000). These scholars generally search for the qualitative equivalents to quantitative approaches to validity, using words such as "trustworthiness," "credibility," "transferability," "dependability," and "confirmability" (Creswell, 1998, 197). In an attempt to move away from quantitative terminology such as "verification" (Creswell, 1998, 197), various alternative words have been used to reference validation, such as the constructivist[5] researchers' call for "authenticity" in research (Lincoln and Guba, 2003, 278). Some poststructuralist[6] or postmodernist scholars call validation "ironic" (Lather, 1993). Other scholars consider validation methods to be "transgressive," questioning truth, reliability, and validity in attempting to bring "ethics and epistemology together" (Lincoln and Guba, 2003, 281). Criticalist researchers are concerned with "advocacy" or the amelioration of oppression in addition to other commonly used validation techniques (Carspecken, 1996). Given the many terms and debates of the validation of qualitative work, it is no wonder then why some scholars have suggested that there is a "crisis in validity" in qualitative inquiry (Gergen and Gergen, 2003, 577). Yet, validity cannot be left in a state of crisis, it must be confronted. Beyond the terminology, there remains a need to somehow speak to the quality of analysis and interpretation in qualitative work. Questions emerge, such as: Do the data provide good evidence for the interpretations made by the researcher? Are participants well represented by the reporting of the study? Was the data analysis technique rigorous? Could the study be replicated?

Qualitative scholars have provided numerous techniques for the validation of their work. Using an objectivist[7] epistemological stance in their analysis of qualitative research in the health sciences, Mays and Pope (2000) assert that relevance is one of the most important techniques of qualitative research validation, indicating that the benchmark for "good research" is whether or not the work adds to knowledge or increases confidence in existing knowledge, or, whether or not the findings can be generalized. Some qualitative scholars maintain that "triangulation," or the use of multiple sources of data, multiple methods of data collection, or multiple investigators, is the most important factor in validating qualitative findings (Denzin, 1978). Others, particularly ethnographic researchers, argue that a researcher must remain in the field for an extended period of time for the work to be considered rigorous or valid (Creswell and Miller, 2000; Fetterman, 1989). Still others suggest that qualitative work must be checked by the participants in the study, a process often called "member checking" (Lincoln and Guba, 1985). Then there are those researchers who maintain that reflexivity, or a researcher's self-reflection, is the most important way to validate qualitative work (Moustakas, 1994). Related to member checking, some scholars conduct a process called peer debriefing, or a research audit, whereby colleagues or people external to the project review the analysis and interpretation (Carspecken, 1996; Lincoln and Guba, 1985).

Among all of these strategies, there is no real standard by which *all* qualitative scholars validate their work. Arguably, imposing such a standard may work against the creative and often status-quo-changing nature of qualitative work. How, then, can research be validated without losing creativity and freedom? How can validation be discussed without marginalizing researchers and research traditions? In particular, what might be the challenges to validity within cross-racial research?

Reflection on the Validation of Cross-racial Research

What makes cross-racial research validation different? Bluntly stated, the possibility for completely missing the proverbial boat when it comes to interpretation is greater when one conducts research with people outside of one's group, regardless of whether that group is a racialized, gendered, religious, sexually oriented, national, or political category. In my view, if all of the validation techniques described above are used, but, the participants are not actively involved in the process, the work may still not be culturally "valid" to the group being studied. That is to say, if one is White, as I am, and one were to only use a White-centric perspective (e.g., White peer debriefers, White research auditors, White reflexivity, etc.), the findings will likely, in turn, be White-centric also. So, how does one journey outside of this conundrum?

I provide ideas here from my own research with African American college women. The study was a critical ethnography of Black women's experiences relative to race and gender in college, examining the way that these experiences influenced their identities (Winkle-Wagner, In Press). In this project, thirty women were separated into eight focus groups, which met biweekly for a period of nine months. I also conducted individual interview and observations during this time.

Perhaps ineloquently, but truthfully, put, I refer to the process of meaningful research and validation across racial lines as "getting real" (Winkle-Wagner, In Press). The research process and findings should be interrogated for colonization qualities. By this, I mean that one of the most important aspects of cross-racial validation is ensuring that there is a match between the participants' analysis and interpretation and the participants' cultural/racial experience so that the researcher is not co-opting the participants' cultural norms, or misinterpreting them. It seems that member checking, the process of asking participants to check the analysis and interpretation of data, might meet this goal. Yet, "getting real" calls for a deeper level of member participation in cross-racial validation. That is, the trust between the researcher and participants needs to be so strong that the participant can feel assured that she can actively disagree with the researcher and that the researcher will use the participant's advice over her own. Preference must be given to the participant. This deeper level of member checking is a bit harder to measure. I knew I had reached this level with the African American women in my study when one of them said to me, "You know Rachelle, you have it all wrong. You really shouldn't be doing it this way." I then asked the participant, who named herself Isis for the study, "How should I be doing it?" Isis responded by saying, "Let me show you." She then took over the facilitation of the data collection for that focus group session and told me how she would interpret it. In my case, the participants became so involved in the project that they in many ways took over the data collection process. I became more of an observer and participant than a facilitator or researcher. The participants began asking their own questions and guiding me through their lives and stories. The participants began facilitating their own focus groups.[8]

The second and equally important technique that I employed was community involvement. While this may seem like part of an ethnographer's tool kit, to become as close to a member in the community of study as possible, I mean more than what some would consider "participant observation." I mean that as a scholar conducting work across color lines, validation of the work can be linked to one's involvement in communities of color. In my case, I had developed friendships and relationships with African American women outside of my research study. These women became valuable challengers to my research and also supportive participants in it. I could ask these women all the tough questions about my own naivety relative to Black experiences. Again, this is not measurable in numerical terms, but this is vital for research validation. These women became my peer debriefers and research companions. They not only read countless transcripts and reanalyzed them to compare their analysis to mine, but also monitored my reporting in order to scrutinize my implicit and explicit racialized interpretations of the participants' stories. The process was transparent to all involved.

The data analysis process is also an important component of validation of work across racial lines. In my study, I conducted a variety of analyses on the data and compared the findings of each of these analyses. For example, I coded the data using low-level codes that were very explicit to connote the broad topic under discussion (e.g., family, friends, classroom issues, racial stereotypes, financial aid, etc.). Then, I used the critical analytic techniques developed by Phil Carspecken (1996), such as meaning field and reconstructive horizon analysis. Meaning field analysis is a way to consider the range of possible meanings behind a statement. Reconstructive horizon analysis deconstructs the data to examine the ontological[9] categories for meaning: subjective, objective, normative, and identity claims.[10] I then, with the help of my peer debriefers, colleagues, and participants, developed my own analysis technique to consider the type of statement made by participants and analyzed (a) to whom the statements referred (i.e., monological—to oneself; dialogical—to others); (b) whether the statement was in the first-, second-, or third-person; and (c) whether the statement was biographical of others or autobiographical. These various data analysis techniques served to validate each other. If there were differences between the findings and interpretations gathered from these techniques, I then delved deeper into the data and discussed the findings with participants and peer debriefers.

Also related to the data, I employed negative case analysis and strip analysis (Carspecken, 1996) as validation techniques. For negative case analysis, I took those parts of the data that did not fit the general themes and trends and reconsidered them. I discussed these portions in particular with my participants, peer debriefers, and other colleagues to determine their meaning and the reason for their differences. To conduct strip analysis, as suggested by Carspecken (1996), I reanalyzed my data in strips away from the full transcripts to see if I interpreted the data in the same way.

Shifting the focus away from the data and participants a bit, I examined my own biases and values. There are many ways to do this, and often researchers call this reflexivity, as I have done above. I did this formally and informally through journaling, field notes, and conversations with colleagues, participants, and people in the larger African American community where I was conducting the research. All together, this created a thorough, self-reflective process whereby I continually questioned my own thoughts about race and my own experiences as related to those of my participants. Additionally, as part of this process of self-reflection, I deliberately educated myself on African American/ Black[11] popular culture (books, music, movies, etc.).

In terms of prolonged engagement in the field, I do think that this can be a validation technique, and it is one that I used in my project. However, I differ from ethnographic researchers in that I think that trust and validation can be ensured without always spending a long period of time in the field. If one is initially close to the participants in some way (e.g., already friends, colleagues, or acquaintances, or already engaged in a community with the participants), for example, one may not need to spend as long in the field in order to collect meaningful and "valid" data.

Finally, implicit in many of the techniques that I recommend for cross-racial research validation is trust.[12] Again, this is a difficult concept to measure, but it is a vital aspect of validation. That is, how can a researcher know that her work is valid if her participants do not trust her? The participants could be holding back information or even trying to give the researcher what he/she desires. If there is a high level of trust, it is more likely the participants will provide honest accounts of their experiences, even if there is a potential for hurting the feelings of the researcher. This is particularly important in cross-racial research because there may be times when the participants will describe experiences where they were hurt by people in the researcher's racialized group. This occurred often in my project. The women described instances where White students, faculty, and staff had overtly discriminated against or been hostile toward them. The women also discussed some of their own negative feelings toward White people in general and White women in particular. At first, the women would often apologize immediately after saying these things. I often would express to the women that these statements did not offend me and that I wanted them to be open and honest. Eventually though, after the women took more ownership over the project and I became more of an observer, the women were very open about cross-racial hostilities and disagreements. Sometimes it is quite difficult to evaluate whether or not a participant is being completely honest about their own feelings toward another racialized group, particularly if the researcher is a part of that group. However, if there is a high level of trust, there is a great possibility for the participant to be frank. In this way, trust between participants and the researcher is an integral aspect of the validation process.

Conclusion

To undertake cross-racial research well and to ensure that this research is valid, it takes a good deal of self-reflection, humility, and acceptance of being wrong much of the time. Common validation techniques could be mostly employed as a checklist for the researcher, and the research could still wrongfully interpret racial issues. If there is a primary point that this chapter highlights, it is that validation of cross-racial research entails engagement, involvement, and community between researchers and participants. Not unlike Supreme Court Justice Potter Stewart's famous definition of obscenity,[13] claiming "I know it when I see it," the participants and those from their racial group will know if the research work has been properly validated. For the researcher, however, it is unfortunate that this judgment often may come too late—after the work is already published. To avoid this pitfall, a researcher cannot be far removed from the racialized groups or communities that she desires to study. It means shedding preconceived notions of neutrality, objectivity, and separation between researchers and participants. It means getting real.

Notes

1. Validation in this chapter refers to the process that a researcher undergoes to ensure that findings and claims in the research are trustworthy, valid, or authentic. Similarly, validity, for the purpose of this chapter, refers to the result of that process.
2. Epistemology is a way of knowing. For example, one using an objectivist epistemology "knows" primarily through the senses—through observation,

hearing, touching, smelling, tasting. In constructivist epistemology, often used in qualitative scholarship, knowledge is constructed in the research process by the participant(s) and researcher, and one "knows" through one's own view of "reality," leading to a multiple realities approach. In subjectivist epistemology, one "knows" through questioning, and knowledge is largely within one's mind or one's personal experience. Finally, in critical epistemology, one knows in relation to an examination of historical and current power structures.

3. Positivists generally use an objectivist epistemological perspective.

4. In the *Qualitative Researcher's Companion,* Maxwell (2002, 1992) asserts that there are different types of validity in qualitative work: (a) descriptive validity: the factual, observable accuracy of a researcher's accounts; (b) interpretative validity: ensuring that the participants' meaning for their experiences is well represented; (c) theoretical validity: the theoretical constructions that a researcher brings to or develops as part of a study; (d) generalizability: the extent to which the research findings can extend to a larger population or group; (e) evaluative validity: the moral/ethical or judgment-based aspects of the study.

5. The vast majority of qualitative researchers align with constructivist epistemology where knowledge is co-constructed as the researcher attempts to understand the reality or world of the participant.

6. Poststructuralists often use a subjectivist epistemological stance where knowledge is within one's own experience or mind.

7. In positivist or objectivist epistemology, the subject and object are clearly differentiated and separated. Alternatively, in constructivist epistemology, the subject and object are linked.

8. For a longer description of the trust-building process within this project, see Winkle-Wagner, In Press.

9. Ontology refers to bodies of knowledge. In this case, the bodies of knowledge are subjective (first person), objective (third person), normative-evaluative (moral/ethical), and identity (what one is saying about oneself).

10. Subjective claims are primarily first-person claims and accessible to oneself. Thoughts, beliefs, and feelings are subjective claims. Objective claims are generally third-person claims and are accessible by multiple people. These are often linked to the senses—something that could be seen, heard, touched, tasted, or smelled. Normative-evaluative claims refer to norms and values. Identity claims are those things that one is saying or implying about oneself. All of these claims are referenced simultaneously, according to Carspecken (1996).

11. I use the terms African American and Black to refer to those with African, Caribbean, or Latino/a-Black heritage. The participants in my study often debated these terms. I use both to represent both sides of this debate.

12. For more information about the process of gaining trust, see Winkle-Wagner, In Press.

13. This famous quote was part of Supreme Court Justice Potter Stewart's dissenting opinion for Jacobellis *v.* Ohio, citing the First and Fourteenth amendment regarding the government's role in censoring pornography, positing that the government was limited in claiming criminality only for "hard-core pornography."

PART III

Exemplars

Engaging the Margins: Working Toward a Methodology of Empowerment

Joshua Hunter

The chapters in this section push the boundaries of research and propel research forward by not only validating marginalized groups but also by further enlisting marginalized and innovative methods. Thus, not only are marginalized groups highlighted and given voice, but the manner in which the scholars approach the subjects stem from marginalized research methods. The exemplars in this section give the reader much to consider regarding the great potential of research that is empowering, contextualized, rigorous, and builds trust between actors within the research process.

As an illustration, Lawrence's piece on American Indian histories not only considers a group who have been marginalized by society and even scholarship, but she does this through the vehicle of micro-histories, a form of scholarship that is innovative, although marginalized, due to its ethnographic, non-generalizable orientation. As she suggests, micro-histories, while not being generalizable, provide a contextual map of issues underlying broad events of cultural, economic, social, and political significance. Positivist research methods and methodologies typically stress the need for generalizability, yet this generalizing often forces marginalization, for it does not account for the great complexities and diverse perspectives inherent in research. Without generalizing, then, this form of historical analysis contextualizes our understanding of where we have come from and what we are influenced by. Micro-histories can therefore articulate idiosyncratic events that allow us to understand both nuanced contextual history and also inform us about ourselves.

If Lawrence's chapter provides us with contextualized historical analysis, Ortloff's chapter provides an example of work that is theoretically substantive, meaning that her theory is informed by her experience in the field, her understanding of social and historical phenomena embedded in the contextualized nature of her research. In examining citizenship education in Germany as a potential vehicle for exclusivity and marginalization, she underscores how teachers, as agents of the state, can either reproduce or resist marginalization. Both Lawrence and Ortloff are asking similar theoretical and historical questions: who belongs, who doesn't? What makes a good citizen (or good Indian)? What role do educators play in this process? And both are articulating historical or theoretical research that is grounded in the complexity of human interactions and idiosyncrasies.

These exemplars provide a glimpse of innovative and potentially profound methodologies for challenging the relationship between the researcher and the researched, bridging the gulf between principle actors in the research process. These examples of action research in methodology and substance excel in explicating not only complexities inherent in marginalized groups but also in postulating that marginalized methods may better attend to research regarding these groups.

This is empowerment methodology; empowering to the researcher for freeing her from the constraints of a positivist straitjacket, and empowering for the people being studied, as it gives them voice, brings them into the research process to tell their own stories in their own way, to share their ideas, and more fully allows for reciprocity and trust. This trust can only be gained through intimate and/or a prolonged research process. While I would agree that length of time in the field is one way of building trust and thereby coming to deeper conclusions, building trust through intimacy and empathy is another. Intimacy is one way of building trust when research does not focus on humans and their interactions, such as Lawrence's, Ortloff's, and Yonehara's (described below) chapters. Intimacy in these examples comes from building awareness of the multifaceted, nuanced, and absolute humanness of the historical events, theory, and quantitative data under investigation. For, after all, the human element, in all its foibles and achievements, its oppression and resistance, even its distance in time or space, is still the central story being depicted. Trust emerges through intimacy and empathy by giving voice and validity to these actors, whether historical or contemporary, through the subjectivity of the scholar. As the first section articulated, denying the subjectivity of the researcher limits what Apple called "the ability to critically engage with the knowledge systems of a community." A critical engagement would require what so many of these examples highlight, innovative application of theory and methods while at the same time privileging the building of relationships of trust between researcher and subject.

The chapter by Yazzie-Mintz asserts that the power of intimacy of knowledge "ensures access to the complex whole of an individual, group, or institution." To get at intimacy, as these different chapters would attest, requires stripping away the apparent gulf that separates researcher from those being researched, thereby maximizing the agency of participants in the research process. The research process, enlisting different goals than objective documentation of experiences, can thereby ensure "the researcher has an opportunity to cultivate intimate biases about the educational context, and these intimate biases contribute important lenses for interpretation of data resulting in the generation of useable and informative research findings. Individuals who hold intimate bias are able to make specialized interpretations and use of local epistemologies, philosophies, and behaviors within the realm of instruction and meaning making." What is important to maintain is that while trust and reciprocity are being constructed, moving away from merely extracting information, they can provide a depth of analysis that is richly authentic and intensely valid. Moving away from strict objectivity does not preclude this style of research from vigorous validity checks. Their analysis is no less rigorous or exacting for creating these relationships; it is purely a change of motivation and goals, perhaps a change of consciousness that occurs as a result.

Payne's chapter examines the role of the life story in exploring lesbian youth identity and postulates an intimate and holistic view of the participants. She suggests that the life story offers an examination of "our own understanding of our past, present and possible future." This correlates with Lawrence's position that by looking at these contextualized historic narratives, we can obtain an understanding of the significance of marginality through time. Payne tells us that the truth inherent in the story lies within the meaning each story holds for the storyteller, within the cultural structures framing the story, and how the story allows navigation of various social situations. Again, this is an innovative method for understanding people and interactions in an intimate and holistic way.

Why then is this style of research often marginalized? Is it the intimacy or the seeming subjectivity that so affronts the dominant research process? Are the lines between researcher and subject too blurred for our positivist orientation? If one of the central questions of most research is an exploration of what it means to be human at a given time and place, then what are we afraid of in validating this vein of research?

It may be that we have much to glean from research that not only explicates the human condition but builds human relationships and community at the same time. The drive toward objectivity forgets that a researcher who is totally emplaced has much to offer and cannot be judged on a natural-science validated scale of objectivity, for the researcher's work and what motivates the researcher is completely different—not invalid, only different.

Even in academic writing we have perpetuated a constricted set of parameters, which disallow either intimacy or artistic creativity. Hunter's chapter not only examines a marginalized group (adolescents) but frames the writing in a creative and engaging way. She discusses the problem of teens dealing with the liminality of mainstream schooling, and it would seem plausible that academic writing also deals with this liminal space, which imprisons researchers from writing in alternative formats that are recognized as equally rigorous. Hunter's work validates adolescent voices as a form of resistance and agency. Her description and interpretation of data evokes this through the creativity of her words. This is the kind of writing that can have an impact on policy and practice, for it has a broader appeal (not solely academic) by revealing the often hidden realms of adolescent sexuality in a creative and accessible way. To effect change, research must attend to the larger audience in ways that appeal to their sensibilities, to their values of equity and fairness, and yes, even to the heart.

Writing for a solely academic audience has left a great deal of research sterile, even when it deals with actual humans who are living actual human lives. What is revealing about these chapters is that the "site" in the research provides great meaning for the researcher, and the writing that comes from this intimacy and deeply felt conviction provides a deeper sense of what is actually transpiring in the data. The "site," (and this should be interpreted broadly to include "historical" sites, "physical" sites, even "bodily" sites) of any research is interconnected to the participants, and a researcher cannot begin to understand the people without understanding the "site." For example, seeing students in school, and watching their interactions in school, is completely different than handing out a survey. Watching their reactions, their networks, their negotiations as they navigate their environment, depends upon intimate awareness of place and how participants interact with and through it. The six chapters in this section give a glimpse of research practice that is intimate, holistic, subversive, challenging, and outright good reading.

The examples in this section broaden the boundaries of research in a way that enriches our understanding, but it goes further to empower the groups or individuals being studied, the researcher and the research achieve something beyond the rigorous interpretation of meaning to build empathy and relationships with the communities they come into contact with. This is not to say that a researcher must simply hang out long enough, or get to know subjects well enough, to achieve exemplary results. All of these examples are rigorous in their methodologies, rigorous in their application of methods, and rigorous in their analysis of the data. But the intimacy born out of ethics, social justice, or basic values of human worth and dignity, drives these studies to new horizons of action research. The margins are marginalized because in part they have no voice and in part they have no support. These exemplars foster the environment necessary for the actualization of voice by supporting and validating these perspectives with innovative and intimate methods.

These exemplars beg the question about what are the goals of research. Are they merely academic? If so, then what positive good comes from them? Is it possible that there are or should be larger, more ephemeral goals than an article, or book, or conference presentation? Should research enliven our discourse, uplift and empower the margins, and engender positive social change? In the end, this is what these six exemplars are suggesting. That there is not only great power in research but perhaps an even greater responsibility to those being studied, to those on the margins of society, to social justice issues, to our interdependent life on the planet.

Yonehara's chapter is a case in point about this larger goal of social justice, for, as she asserts, her model attempts to "make 'the invisible people,' such as rural children who are often ignored in policymaking process, more visible." Her quantitative work enlists human well-being theory to articulate the premise of education as a human right and furthers this idea with work that helps to contextualize literacy issues in Tanzania. This is quantitative research at its best, delving into a world in which real people live real lives and real oppression occurs. The attempt is to articulate how this type of data (which is typically vacant of human reality) can begin to attend to these contextualized issues. A powerful way that Yonehara achieves this type of intimacy is her prolonged time in the field (over a year) and her prolonged period of living with the data during analysis (over a year). She is able to examine the human element in her numeric data because she was able to intimately connect in the field and in her analysis.

It is time that we examine marginality in all its many facets, marginalized groups and individuals, as well as marginalized theories and methods. Perhaps it is the case that only by investing in alternative and historically marginalized methods that marginalized groups can be afforded a voice in the research process. If this is the case, then these examples from the field, from the margins of a world of research still held to positivist and objective norms, may be a key to not only explicating the complexities of social interactions, but also empowering marginalized groups and the research itself to speak from unique perspectives.

Only by truly investing in humanity and building trust and relationships of reciprocity and altruism will we begin to discern humanity in a clearer light. This will not only make for more authentic and appropriate research, but force all of us in positions of power to consider the human element in our decision-making. Thus, this is the great potential for this type of empowerment research, giving agency to marginalized groups and giving validation to marginalized methods to challenge the status quo of mainstream research, which disavow intimacy and the building of trust and community.

This section is not only informative about innovative research methods but also asks the reader to challenge his/her own assumptions about knowledge acquisition, the role of research in bettering human society, and the reliance upon the strengths and perspectives of the subjects in a study. Each one stands alone as a unique application of marginalized methods of research, be it micro-history, life histories, portraiture, or innovative ways of documenting and reporting data and challenging theoretical ideas such as human well-being theory or critical ethnography. Taken together, they are a strong foundation for moving themselves as researchers, and the academy as a principle audience, toward an alternative path of research methodology, methods, substance, motivations, and outcomes.

CHAPTER 9

Educating the "Savage" and "Civilized": Santa Clara Pueblo Indians at the 1904 St. Louis Expo

Adrea Lawrence

On Microhistory and Education

A microhistory is history on a very small scale. The unit of analysis tends to focus in on a particular event or place, a specific sliver of time, a distinct group of people, or some combination of these. Historian Edward Muir (1991) writes that by honing in on the particular, the microhistorian can create an "ethnographic history" (ix) in which s/he attempts to understand past people as they experienced the world by weaving together a contextual tapestry (Revel, 1995). Microhistories have a prismatic effect, refracting larger social phenomena through the contexts and experiences of the people studied. Microhistories can inform us of attitudes, tensions, and motivations underlying broad social, political, cultural, and economic phenomena even though microhistories are not generalizable.

This chapter is a brief and partial microhistory of the 1904 Louisiana Purchase Exposition in St. Louis, Missouri, otherwise known as the St. Louis Expo. Like the 1893 Columbian Exposition in Chicago, the Bureau of Indian Affairs (BIA) organized exhibits calling attention to the U.S. government's efforts to assimilate and formally school American Indian[1] children (Trennert, 1987). Indeed, the whole 1904 St. Louis Expo was billed as a demonstration of "educational progress" ("Educational Progress of the Year," 1904) and as an "educative force" (Slocum, 1904). Particular attention was to be paid to the century of developments in the geographic area that was part of the 1803 Louisiana Purchase ("The St. Louis Fair," 1903). One could argue that the primary purpose of this event was to trace how North America became the United States through the colonization of much of the continent. A century's worth of lessons learned were to be demonstrated, and the Fair itself was to be an educational event for those who attended. Spectacles of technological innovation and exhibit halls featuring the distinctive art, architecture, food, and customs of other countries were showcased. In the early months of 1903, Ernest Hamlin Abbott (1903) wrote that the underlying questions of the Fair were: "What will it tell concerning the character of America? [W]hat will it reveal of the soul of the Nation?" (560). We might add: What does it mean to learn and live in a radically changed and changing environment? None of these questions was limited to the perspectives of European-Americans[2] who might be identified as colonizers of the Americas. Expo participants ranged from scientists from all parts of the globe, to cultural representatives in the several international exhibit halls, to aboriginal peoples from North America. What do we make, then, of those Indigenous peoples who attended the Fair? And, how did their interactions in their home settings influence their experiences at the Expo?

The Players

This microhistorical portrait stems from the correspondence between a BIA day school teacher, Clara D. True, and her supervisor, C. J. Crandall. Their letters offer glimpses into the processes of selecting Pueblo Indians to participate in the St. Louis Expo as well as into the motivations the Santa Clara Pueblo Indians had for attending the Fair. These letters, likewise, convey qualities True and Crandall appear to have believed made a "good" Indian.

Clara D. True, an Anglo woman from Kentucky, served as the Bureau of Indian Affairs Day-School[3] teacher at Santa Clara Pueblo, sometimes referred to as "the Pueblo," in the Rio Grande Valley of northern New Mexico from 1902 to 1907. True lived with her mother at the Pueblo and, although her primary purpose was to teach Santa Clara children, she often functioned as an intermediary between Pueblo[4] community members and her supervisor, Mr. Crandall.

C. J. Crandall, who was also Anglo, worked as the superintendent of the Santa Fe Indian School and functioned as the acting Indian agent for the Northern Pueblos District.[5] Within the organization of the BIA, Crandall had the authority to make decisions about whether or not children at the Santa Fe Indian School could go home for holidays and who was eligible to attend the St. Louis Expo.

Samuel McCowan, the superintendent of the Chilocco Indian School in Oklahoma, was in charge of overseeing and recruiting for the two BIA exhibits at the 1904 St. Louis Fair.

Pedro and Genevieve Cajete, a father and daughter from Santa Clara Pueblo, had known True since her arrival at the Pueblo in August, 1902. Pedro Cajete appears to have interacted frequently with True, in part because the Santa Clara Day School sat on his family's land. The BIA rented the school from him.

The woman potter was also a Santa Clara community member recognized inside and outside of the Pueblo as a skilled artisan.

Making the "Savage" Indian

Writing on May 14, 1903, Samuel McCowan of the Chilocco Indian School in Oklahoma solicited American Indian child participants from other BIA school superintendents for the 1904 Louisiana Purchase Exposition. McCowan (1903) wrote to Crandall:

> I have been appointed "Charge Des Affaires" of the Indian exhibit at St. Louis next year. It is our purpose to maintain a model school there. I hope to have not less than one hundred pupils in attendance taken from the various schools in the Louisiana Purchase Territory. This brings you in. I want to have with us an Indian school band of not less than forty pieces. This band will not be a Chilocco band, a Haskell band or a Santa Fe band but an Indian school band, and I want to collect for this aggregation a few of the best players from the best bands at the various Indian schools. What can you furnish us from you [sic] band? Please give me the names of the boys, their ages, and band experience. I desire to have as many full-bloods as possible.

The Bureau of Indian Affairs asked McCowan to organize two types of Indian exhibits. One featured a model school that displayed Native students in European-American dress in a European-American classroom. The other presented what might be construed as "old" Indian ways of living, which included different types of tribal housing, art, handicrafts, and performance. According to historian Robert A. Trennert, Jr. (1987), the BIA was intent on promoting the success of the Indian School Service in assimilating Native children through a live demonstration of schooling, which stood in marked contrast to the "old" Indian exhibit featuring traditional tribal practices. The tandem exhibits of the model school and the "old" Indian were presented to highlight the differences between the civilized and the savage.

McCowan's letter to Crandall suggests that Indian students would both be on display and would demonstrate their capacity—despite their Native blood—to become more like European-Americans. As the superintendent of the Santa Fe Indian School, Crandall had the authority to decide which Pueblo Indians would attend the St. Louis Fair to represent not only their communities but also the Northern Pueblos District under his supervision.

Between January and February of 1904, Crandall asked True to recruit skilled potters[6] for the Expo. On January 20, 1904, True wrote, "The best potter in Santa Clara Pueblo is willing to go to the St. Louis Exposition" (True, 1904a). True believed the woman's willingness to go to the Expo stemmed from her husband's existing agreement to participate in the independent Cliff Dweller's exhibit, an exhibit upon which the BIA frowned. The woman said she would attend, however, only if she was allowed to take her daughter to St. Louis or enroll her in the Santa Fe Indian School. Crandall gave the woman permission to do either. He indicated that her representation of the Pueblo, as a potter, was important to him; thus, recognizing her bargaining power as a skilled, respected artisan, despite the protocol of her having to request permission to travel from a BIA official. Crandall (1904a) also advised the woman to begin collecting the materials she would need for several months of pottery-making in St. Louis. Throughout February and March, Crandall regularly inquired about the potter's willingness to participate in the Indian exhibit at the Expo, suggesting that he wanted to ensure her presence there. During the same period of time, the potter regularly asked about the transportation to St. Louis and other preparations that needed to be made (Crandall, 1904b; True, 1904b, 1904c).

Crandall almost did not send Indians from the Northern Pueblos District to participate in the government program at the Expo, though. In a letter to True at the end of February 1904, he expressed concerns about McCowan's treatment of the Indians involved in the exhibit. He wrote, "Mr [sic] McCowan expects to secure Indians without money and without price,—that they will construct their own habitations when reaching St. Louis, and live like savages rather than progressive citizens in decent habitations, like most of our pueblos do live" (Crandall, 1904c). Crandall further stated that he would not send people from his district unless McCowan provided for the participants in some way. Although the correspondence between Crandall and True does not indicate so, Crandall apparently came to an understanding with McCowan about the housing for the Pueblo Indians, as the potter and several others from Santa Clara did, in fact, go to St. Louis.

Based on the apprehension Crandall expressed in his letter and Trennert's account of the 1904 Fair, McCowan was responsible for highlighting the contrast between the barbaric and the civilized. From McCowan's letter, Crandall's response, and Trennert's assessment of the Expo, the "old" Indian exhibits, housed in the government building, appear to have been read as static presentations of the exotic, untamed, and primitive Indian. This stagnant display stands in sharp relief to the model school exhibit where the students were transformed into dynamic, progressive, and individualized people who behaved in ways familiar to European-American Expo attendees. The Pueblo Indians, according to Crandall, were civilized Indians because of how they lived. The savage-civilized paradigm simply did not apply to Pueblo Indians.

Recognizing the "Civilized" and "Good" Pueblo Indians

Legally, Pueblo Indians were not considered to be like other Indians in the nineteenth and early twentieth centuries. When the U.S. government concluded its war with Mexico in 1848, acquiring what is now identified as the Southwest, Pueblo Indians were regarded as full U.S. citizens. Through a series of court cases and federal provisions in the early years of the twentieth century, the Pueblos' status as full citizens was progressively pruned back to that of partial citizens and then as wards of the government (*U.S. v. Felipe Sandoval*, 1913). Other Indigenous groups typically followed the opposite trajectory, moving from the status of ward to that of full citizen, which was frequently done through the land allotment process. The rationale for Pueblo Indians' initial full citizen status was based on how they lived—in permanent adobe houses as agriculturalists who practiced a hybridized version of Catholicism (*United States of America v. José Juan Lucero*, 1869).

Although Pueblo communities lived in ways that appeared familiar and "civilized" to European-Americans, European-Americans simultaneously recognized that Pueblo communities comprised "Indians." This does not mean that Pueblo Indians were themselves unfamiliar with European-American attitudes and practices. The fact that Pueblo Indians did attend and participate in the St. Louis Expo suggests a double consciousness (DuBois, 1903): they knew how to present themselves as Indians in ways Crandall thought were civilized and they understood the sociocultural roles and expectations of behavior within their own communities.

Securing the participation of Indians whom True and McCowan deemed as "good" is prominent in their letters. McCowan (1903) wanted to feature "the best players from the best bands" to play in the band he created for the Expo. Having the "best" or "good" would seem to extend to Indians going to the fair in other capacities. When Crandall asked True to inquire about other skilled potters who would be willing to attend, True gave her assessment of potential participants' skills and moral character. She wrote, "There are plenty of women who would go but they are not needed here nor elsewhere, as they are idle, dirty loafers who wouldn't make pottery worth the name" (True, 1904b). True's statement implies that the quality of one's work somehow reflects on the quality of one's moral character. This might reveal her own standards or those of Santa Clara. If the women were indeed loafers, as True asserted, it is unlikely that the Pueblo would have recognized them as prominent community members. Anthropologist W. W. Hill (1982) observed that the virtue of industry was one of the central values in Santa Clara society, suggesting that if the people of Santa Clara did not perceive the women True described as industrious, it seems doubtful that the women would have taught pottery to children.

True also lobbied for individuals she felt worthy of participating in the St. Louis Expo. Through True, Pedro Cajete, who rented the Day School property to the U.S. government, asked Crandall if he and his daughter, Genevieve, could attend the fair. True reported that Cajete had already obtained permission from St. Catherine's, the nearby Catholic school, for Genevieve to attend the Expo. True added that although Cajete had initially intended to go to the fair with the Cliff Dwellers exhibit, which was not sponsored by the BIA, he ultimately did not trust them and preferred to go as a participant in the government's Indian exhibit. True argued in favor of Cajete and his daughter, referring to Cajete's cooperation with Crandall over the years and Genevieve's appearance and character. True wrote:

> Genevieve is rather nice looking and makes passable pottery. She has any amount of handsome Indian clothing which she would expect to wear exclusively. Her father showed me the wardrobe today. It is very nice from an Indian point of view. Pedro has numerous warlike garments. He will lay in a supply of native products, silver jewelry and all sorts of similar articles. (True, 1904d)

The Cajetes were "good" Indians in True's eyes. It should be noted that they negotiated successfully between her world and expectations and those of the Santa Clara Pueblo. Pedro Cajete seemed to understand that appearing Indian was a fundamental qualification to attend the Expo. He demonstrated this by showing True both his and Genevieve's Indian clothing and conveying that he knew when it was appropriate for him and Genevieve to be perceived as Indian. In doing so, he was uncompromisingly polite, which seems to have further endeared him to True. This is evidenced as True referred to him by his first name throughout the letter. In the collection of her letters, she appears to have called individuals by their first names only when she was on friendly or familiar terms with them. True closed the above letter by saying that the Cajetes would complement the work of the Souseas, another Santa Clara family, who were already set to go to the Expo. She ended with further evidence of Pedro Cajete's desire to attend and his capacity to function as a diplomatic and intelligent representative: "If you don't let him go, I'll have him to bury and he is so shrewd he will beat me out of his funeral expenses" (True, 1904d). Crandall accepted True's evaluation, and the Cajetes attended the Fair.

True, Crandall, and Cajete all seemed to anticipate the draw of the appearance of the exotic Indian. Serving as gatekeepers and regulating who could and could not participate in the government's exhibit at the St. Louis Expo, True and Crandall determined who could be a government sanctioned Indian in front of an audience of non-Indians. Those for whom True advocated were all "good" Indians; she felt comfortable enough in her interactions with and her perceptions of them as valuable community members to recommend them to Crandall. In other words, those from *American Student Achievement Institute* Santa Clara who participated in the government exhibit knew how to interact with non-Indians in syncretic ways that would allow them to maintain their own societies.

The Adventurous, Entrepreneurial Exotic

The St. Louis Exposition itself was a grand exhibition of contradictory ideas about civilization and American Indians. Through exhibition pamphlets and letters from government officials in charge of the Indian school demonstrations, Trennert (1987) maintains that from 1893, when the U.S. government began sponsoring BIA exhibits, to 1904, when it participated in the St. Louis Expo—its last world's fair—fairgoers were drawn to exhibits that depicted Indigenous groups as exotic. Juxtaposing Indians in their traditional regalia and engaged in tribal rituals, with Indians dressed as European-Americans in classrooms, did not seem to convince attendees that European-American social values and practices would or should triumph over Native cultures.

Being Indian—appearing Indian and acting Indian—were the well-attended curiosities of the Fair.

Trennert (1987) notes that the exotic, traditional Indian exhibits and demonstrations were among the favorites of the Expo, thwarting the BIA's goal of displaying assimilation as a compelling sign of social progress. In fact, when BIA superintendents were approached to recruit "old" Indians for the Expo, several responded that nobody was available because assimilation had been so successful ("Correspondence," 1903). Whether or not these superintendents saw what they wanted, or what they were meant to see, is unclear. However, their assumption that American Indians were not interested in going to the Fair did not hold for every Indigenous North American group. In a letter to *Outlook* magazine, one Indian agent quoted from the catalogue of an American Indian farmer who produced traditional handicrafts:

> All goods quoted are genuine Indian design and made by them in the old way. We allow no thread to be used by them in the construction, but insist on sinew being used exclusively, as that was the only thread known before the white man came and is much more lasting. With very few exceptions, all the articles quoted are practical. Take, for instance, the golf-belts; they are original and beautiful, practical and durable; combined with the purse there is no more practical article in the market. So it is with moccasins and card-cases, music-rolls, book-covers, etc. ("Correspondence," 1903)

The American Indians who created the golf belts and purses for non-Natives clearly accommodated the demands of buyers who likely lived beyond the tribe's reservation. The assumption that Indians were not savvy about the presentation of themselves and their goods also did not hold.

For members of the Santa Clara Pueblo who attended and participated in the St. Louis Expo, the opportunity to make money and perhaps travel and interact with a variety of different people appears to be significant. As suggested above with the potter's husband and with Pedro Cajete, Pueblo members had planned on attending the fair with the Cliff Dwellers exhibit before opportunities arose with the government exhibit (True, 1904a). Although True and Crandall's letters do not show any indication, Trennert (1987) states that the Cliff Dwellers exhibit, which staged tribal dances and an Indian Congress, was put together in order to turn a profit. The BIA exhibits were not. Although profit was not a stated intent, participants in the BIA's Indian exhibit could sell the goods they produced. In his letter to True, on January 22,

1904, Crandall remarked that the potter from Santa Clara should be able to make money on the pottery she made and sold at the Expo (Crandall, 1904a). Likewise, Pedro Cajete had assembled a variety of Indian handicrafts to take with him and presumably sell at the Expo (True, 1904d). Considering that this may have been one of the few ways the people of Santa Clara could make money quickly, the opportunity to sell goods to people who had never seen Puebloan pottery, jewelry, beadwork, and blankets might have been very enticing.

Pedro Cajete's status as a "good" Indian became something of a conundrum for Crandall as Cajete expressed his favor for an Indian school beyond New Mexico. After their arrival in St. Louis in mid-May, True wrote that Pedro Cajete said he was "having the time of his life" (True, 1904e). Cajete also seemed to get on well with Superintendent McCowan while at the Expo. Upon his return to Santa Clara, Cajete demonstrated his support for the Chilocco Indian School in Oklahoma, which McCowan ran, by recruiting Pueblo Indian students on its behalf. It appears that Crandall thought Cajete and McCowan may have been in cahoots. Crandall sent a letter to Cajete regarding the student recruiting:

> I have no objection to your soliciting pupils in the pueblos [for Chilocco], provided that you do not entice any of my pupils [who attended the Santa Fe Indian School] who are spending the summer vacation at home, to go with you. You are further told that you must take no Pueblos from Santa Clara or any of the other pueblos without first bringing them to this office for my approval and the doctor's examination. If you should di[s]obey this order i [*sic*] you will get your self in trouble with this office. (Crandall, 1904d)

In recruiting students for the Chilocco Indian School—in support-ing federal Indian schooling—Cajete had stepped onto the turf over which BIA school superintendents fought. Student enrollment was a close contest in which BIA officials competed with each other and indi-vidual Native parents; Crandall did not want to lose students to McCowan. Despite the earlier commendations of Cajete's cooperation with U.S. government officials, Indians were not to co-opt federal pol-icies on their own. By invoking his procedural authority over Cajete, Crandall gave the appearance of muting Cajete's personal feelings, as well as his own, for either the Santa Fe or the Chilocco Indian schools, and adhering to the rules, regulations, and hierarchy the BIA demanded. Such a move might have been appreciated by Crandall's superiors and might have made him appear impartial and consistent among the Pueblos. In other words, Crandall might have found and legitimized his authority by following procedure correctly and requiring those supposedly under him to do the same.

Lessons from the Expo

What do these experiences surrounding the 1904 St. Louis Expo tell us about "the character of America," "the soul of the Nation," and what it means to live in a radically changed and changing environment? Writing several years after the Expo, from her position as the Superintendent of the Morongo Reservation in southern California, True (1909) publicly reflected in *Outlook* magazine:

> It is to be regretted that in dealing with the Indian we have not regarded him worth while [*sic*] until it is too late to enrich our literature and traditions with the contribution he could so easily have made. We have regarded him as a thing to be robbed and converted rather than as a being with intellect, sensibilities, and will, all highly developed, the development being on different lines from our own only as necessity dictated. The continent was his college. The slothful student was expelled from it by President Nature. Physically, mentally, and morally, the North American Indian before his degradation at our hands was a man whom his descendants need not despise. (336)

True appears to advocate for the longstanding historical learnings of American Indians on their own terms as they emerged within North American environments. She implies that in the rush to colonize or convert American Indians to the ideals and day-to-day practices of European-Americans, valuable Indigenous cultural phenomena were irrevocably lost. The ancient learnings of Indigenous peoples across the North American continent was subsumed in the rapid expansion of non-Native settlement and disregarded in United States government policies to remake Indians in the social, political, economic, agricultural, and educational image of European—preferably Anglo—Americans. Yet, True, like Crandall and McCowan, was complicit in this process.

One of the central tensions we see manifested is the dichotomous thinking that McCowan, Crandall, and True all expressed to some degree. Whether or not McCowan could get the "best" "full-bloods" for his band, Crandall could squelch Cajete's student-recruiting efforts, or True could judge character based on skill as a potter, each of these BIA employees operated on some notion of what a "good"—and by extension "bad"—Indian was. When these notions were challenged, as with the large attendance at non-BIA Indian exhibits, there was limited space for a reflective response. But, as True suspects in the quote above, American Indians had something to teach—how to read the Other, learn to see the Other's position, and negotiate in mutually intelligible ways.

Notes

1. The term "American Indian" refers to aboriginal peoples living in North America prior to European settlement. American Indian, Native, and Indian are used interchangeably here. When appropriate, specific tribal affiliations are used.

2. European-American refers to those people who immigrated or whose ancestors immigrated to North America from European countries. Racial and ethnic identification in New Mexico in the nineteenth and twentieth centuries is complex, reflecting waves of colonization. Spanish colonists, who settled in the region from the sixteenth through the early nineteenth centuries, could certainly be denoted as European-Americans. In the nineteenth and twentieth centuries, members of this group frequently referred to themselves as "Spanish" or "Hispano," even if there had been intermarriage with Indigenous peoples in the area (Nieto-Phillips, 2004) and were members of different social classes. When the United States acquired the territory under the Treaty of Guadalupe Hidalgo (1848), Hispanos were legally regarded as "White," despite disparate social, economic, and political treatment. The term "Anglo," therefore, offers more precision when referring to U.S. newcomers in the area who also would have been regarded as "White."

3. Bureau of Indian Affairs day schools were located on tribal or reservation lands within Native communities. Day schools were designated for children under twelve years of age. Boarding schools were established for children over twelve or who lived in communities where no day school was present. Both the Santa Fe Indian School and the Chilocco Indian School were off-reservation boarding schools.

4. Today, there are nineteen Pueblo Indian communities that have been situated as sedentary agricultural settlements along the Rio Grande and its tributaries for centuries. Pueblo Indian societies are unique and are typically identified by their geographic, linguistic, and sociocultural characteristics. Santa Clara Pueblo is one such community located about twenty-five miles northwest of Santa Fe.

5. The Northern Pueblos District was created by default when the Pueblo and Jicarilla Apache Agency was dissolved in 1900. The superintendent of the Santa Fe Indian School was to supervise the Pueblo day schools, including that of Santa Clara. This arrangement lasted until 1911 when the position of Superintendent of Pueblo Day Schools was created (Svenningsen, 1980).
6. Santa Clara Pueblo is known for its black, high gloss pottery.

CHAPTER 10

Intimately Biased: Creating Purposeful Research in American Indian Education with Appropriate and Authentic Methodology

Tarajean Yazzie-Mintz

Western scholarship about Native peoples continues to dominate published works, academic departments, and projects directed from within universities and research institutions across the United States. It is clear that Western scholars have a defined purpose and research agenda. But the articulated purposes are distinct from what contemporary Native researchers and communities are proposing. I argue in this chapter that (1) all research is purposeful, and (2) purposeful research in American Indian education must be matched with appropriate and authentic research methodologies that respond uniquely to Native-defined purposes and needs.

In their renowned study of the Navajo, anthropologists Clyde Kluckhohn and Dorothea Leighton (1974) carried out a research agenda guided and powerfully shaped by work for the Indian Service Administration, a branch of the U.S. government dedicated to dealing with Indian affairs, which was focused on "changing" the ways of the Navajo people. Their book, *The Navaho*, was envisioned as a guide to embark on this social and cultural change project, the object of which was the Navajo people: "...too often administrators have forgotten that to change a way of life you must change people, that before you can change people you must understand how they have come to be as they are" (Kluckhohn and Leighton, 1974, 27). To generate psychological and cultural transformation of the Navajo people, these authors must begin by understanding "how they have come to be as they are." Kluckhohn and Leighton's discourse demonstrates the paternalistic nature of many early scholarly works focused on Native peoples. Their research question, "How can minority peoples, especially those which have a manner of life very different from that of the Euro-American tradition, be dealt with in such a way that they will not be a perpetual problem to more powerful governing states?" (Kluckhohn and Leighton, 1974, 25), signals their purpose: studying "minority peoples" as a "perpetual problem" to "be dealt with." This research agenda reflects dominant positions of power and perspective, and within this dominant paradigm, Navajo people are relegated to units of analysis rather than actors in beneficial inquiry, objects of study to be changed for the government's purposes rather than a community of people with agency.

Early inquiries, such as those of Kluckhohn and Leighton (1974), represent purposeful research about Native people in which all aspects of the research methodology—questions, entrance, observation protocols, interview processes and structure, use of recording devices—were defined from the outside in, from non-Native ways of thinking and purporting "objective" truth. Research on Native communities by distant outsiders was, and continues to be, perceived, despite clear evidence to the contrary, as "objective," "distant," and "unbiased," in terms of both analysis and research methodology.

Vine Deloria, Jr. (2004) traces the transformation in purpose from research and scholarship conducted by non-Native scholars to the visible scholarship of Native researchers documenting their own histories, historicizing the evolution of an alternative philosophy, a philosophy that places Native scholars and Indian communities, as a whole, at the center of intellectual and scholarly inquiry. Despite this difference in purpose, Deloria asserts that it was unlikely that the first generation of Indian scholars viewed their scholarship as "dedicated to making Indians visible in significant numbers in intellectual circles. Nor did they consider that the tribal traditions they wrote and spoke about represented an alternative philosophy to Western materialism" (16). Deloria indicates their purpose was only to present accurate information about Native peoples, but it did not include making attempts to change the ways in which the academy operates or how others might conduct research about Native people and communities. Contemporary Native intellectuals, such as Devon A. Mihesuah (2000), are attempting to change the ways in which research on Native communities is conducted and used, asking important questions such as, "Where is the information anthropologists are supposed to be acquiring that can help present day tribes?" (96). Scholars of Native studies have raised critical questions about the usefulness of research to Natives studied and written about by researchers (Greymorning, 2004; LaFromboise and Plake, 1983; McCarty, 2002; Mihesuah and Wilson, 2004; Smith, 1999; Swisher, 1998; Wilson, 1998, 2004; Yazzie, 2001, 2002). Contemporary ethnographic studies of schooling have set a new measure of research, where the researcher frames inquiries of teaching and learning with teachers, students, administrators, and local community (for examples, see the works of Lipka, 1998; McCarty, 2002; Smith, 1999).

Julie Kaomea (2004), a Native Hawaiian scholar of elementary and early childhood curricula, explains the complexity of identifying both as a member of the Hawaiian community and as an academic conducting the inquiry, in a context in which the former position places her as an insider and the latter position is conceptualized by Native Hawaiians as an outsider. Kaomea agrees with Smith (1999), a Maori scholar, that "there remains a very real ambivalence in Indigenous communities toward the role of western education and those who have been educated in western universities" (Kaomea, 2004, 27), particularly when Indigenous scholars run the risk of being socialized and trained to conduct research that will not be responsive to Native communities' needs. Educational researchers who are concerned about the value of research to the community being studied purposefully employ a methodology to seek answers to authentic questions that are defined by both those who are studied and those who are conducting the inquiry. Kaomea and Smith provide reason to question in what ways might the research process and findings contribute to innovative change or practices?

The Lakota scholar, Angela Cavender Wilson (2004), contributes a rational justification for the central involvement of Indigenous scholars in conducting research with Native communities to achieve particular goals and develop and sustain Native existence, identity, survival, and resistance:

> . . . we can best be of service to our nations by recovering the traditions that have been assaulted to near-extinction. For us, our traditions provide a potential basis for restoring health and dignity to our future generations. (69)

These contemporary research purposes—to contribute to bettering circumstances, to recover Indigenous traditions, and to restore health and dignity—raise methodological challenges that force an unveiling of ideological and intellectual location of researchers and ultimately make public, within the academy, debates about accepted objective/subjective scholarship. Wilson (2004) believes that an ideological and political location is necessary in order to counter "those who have been defining our existence for us and who have attempted to make us believe we are incapable of self-determination" (74).

Native researchers and few non-Native scholars are the witnesses, recipients, and holders of essential tribal knowledge, which cultivates an "intimate bias." The holding of intimate bias, having knowledge about practices, traditions, and educational and research spaces, ensures access to the complex whole of an individual, group, or institution. Lawrence-Lightfoot (1997) indicates that attempting to maintain distance so as to be "objective" can minimize or "undermine productive inquiry" by "minimizing their [participants'] authority and potentially masking their knowledge" (137). Individuals who hold intimate bias are able to make specialized interpretations and make use of local epistemologies, philosophies, and behaviors within the realm of instruction and meaning making.

Teachers hold intimate biases, utilizing this knowledge to make decisions about how and what to teach, and what Native knowledge can be incorporated into schooling. Intimate biases help Indigenous scholars firmly locate the activity of research within contemporary, historical, and living Indigenous epistemology and ways of being. The holding of intimate bias contributes to the generation of firm commitments to tribal nations as a first audience, and then to the academy as a secondary audience. The researcher has an opportunity to cultivate intimate biases about the educational context, and these intimate biases contribute important lenses for interpretation of data resulting in the generation of useable and informative research findings.

Much debate exists within research literature about the effectiveness of Native versus non-Native researchers in Native communities, particularly in regard to insider/outsider positionalities. However, these debates accomplish little in moving us toward conducting research that is attentive to the needs and purposes of Native communities. We must focus more on how the research will be done, than on who is doing the research, by seeking a methodology that values Native communities as well as the intimate biases of individuals by asking the following questions: How do research questions from inside communities shape research methodology that is responsive to the needs identified by the community? What intimate knowledge helps Native researchers investigate and document the work of Native educators in authentic and purposeful ways? Can educational research about Native people by Native people influence Western academic and methodological discourse and action? Why is there a need to consider decolonizing research methodologies in Indigenous educational contexts?

In the next section I discuss the ways in which Portraiture, a qualitative methodology, enhances the possibility of engaging in purposeful and meaningful inquiry with Native individuals and communities. Likewise, Portraiture offers great potential for speaking simultaneously with multiple audiences (i.e., Native communities, generalist educators, and the academy) through individual portraits of teacher practice.

Portraiture as an Authentic and Appropriate Methodology

Portraiture, a qualitative methodology framed by the phenomenological paradigm and ethnography developed by Lawrence-Lightfoot and Davis (Lawrence-Lightfoot, 1983; Lawrence-Lightfoot and Davis, 1997) is useful in documenting Native education, particularly the work of teachers' cultural and linguistic knowledge used in Native educational contexts. Through Portraiture, a researcher constructs a narrative picture of the world through the participants' eyes. In my work with teachers in various linguistic and cultural contexts, I continue to draw on Portraiture as a systematic qualitative methodology, one that allows me to speak with audiences beyond the academy, particularly American Indian communities. Through "portraits," or in-depth narrative "pictures" of teachers, I invite educators, parents, and communities to learn and speak about education in spaces other than universities and academic conferences. It is my hope that by engaging communities in accessible discussions of teaching, Native people can participate in shaping their own social, cultural, and political transformation (Yazzie, 2002).

Using Portraiture, the Portraitist is guided by systematic research protocol that serves to forefront the process of building relationship by developing symmetry, reciprocity, and boundary negotiation with the actors. Also integral to Portraiture research is searching for goodness, and creating an aesthetic whole—the portrait layered with detail and complexity—written to engage multiple audiences in the discussion of new knowledge (Lawrence-Lightfoot, 1997). In the next sections, I outline essential aspects of Portraiture that make it particularly useful for conducting research with Native teachers and communities: searching for goodness, forging relationship and trust, cultivating agency, defining purposeful inquiries beyond the academy, and honoring intimate biases held by both researcher and actors in research inquiry.

Symmetry and Searching for Goodness

In my work with Native teachers, I explicitly state my purpose and intent to seek goodness as a primary purpose of my research contribution. Goodness is not a perfected state of education but rather a holistic portrait that offers complexity to balance stereotypes, offers strengths to balance weakness and offers a healthy view from inside to counter perceptions of pathology from the outside. (Yazzie, 2002, 37)

In my search for goodness, I recognize that Native teachers enter educational contexts with purpose and intimate biases about knowledge, instruction, students, and learning; it is equally important to recognize that these notions and practices are neither perfected nor static. In using Portraiture, I am interested in searching for the complexity in relationships and development of teacher knowledge—not simply an "idealized portrayal" of Native teachers.

Forging Relationship and Trust

As a part of an ongoing qualitative research study documenting the developing language immersion practices, I was invited to work with teachers at the Cherokee Nation. In this study, I observed and interviewed seven language immersion teachers who work with young children enrolled in the pre-K to second grade program, seeking to understand both the challenges and supports that help illuminate goodness, strengths balancing weaknesses, in language immersion teacher practice. The field note below describes a teacher's (Karen) struggle with a three-year-old boy, Triston, during naptime (pseudonyms are used in this text). Also in this section, I highlight how relationship and trust surface the opportunity to see new possibilities in instruction. Below, Karen unveils her own instructional questions that become important to her work as a Cherokee early childhood language immersion teacher:

It was difficult for [Karen] to have me observe while she struggled with Triston. Each time she attended to him she would walk over to me to whisper to me what she was trying to achieve. She also shared her growing frustration each time he got off his cot to roll on the floor, or crawl over to another child to bother that child. [Karen] attempted to continue being very controlled in her engagement with Triston. Her frustration built and she seemed to find comfort in our conversation about how to address Triston's non-compliance with naptime routine in another way. I knew she did not want to lose patience with him. She believed that her only other option would be to use English with the child because he did not speak or understand Cherokee. She also wondered whether "he might be ADD."...She indicated that she believed that in addition to not understanding Cherokee he had little routine [and structure] at home. Both of these theories could possibly help explain his behavior. She asked me what I would do to help him remain on his cot during naptime. We discussed some potential options: Have him engage in a quiet activity like look at a book or work on a puzzle at the table. Another option would be to help him lie down to relax; rubbing his back and staying with him at the beginning minutes of naptime a few times a week could pay off in the long run. Or sit with him speaking in Cherokee or singing a song to calm him down. If she can continue to remind herself that he is three years old and is new to the immersion setting she may be better prepared to find many new ways to connect with him. I was excited to see how she was willing to share her frustration and to ask for my thoughts about what she could try. Asking me about what she might try demonstrated to me that she had developed a sense of trust about me being there witnessing her work. Having me observe her work during a tough time is not easy. She showed strength even during a time she might have felt she was exposing weakness. (Yazzie-Mintz, Field Notes, October 9, 2007)

I knew Karen did not want to lose patience with Triston. She questioned full immersion practices, considering using English with the child because he did not speak or understand Cherokee. She also wondered whether "he might be ADD," a possible theory that might shape her next step to help him learn classroom routine. She wondered out loud about the influence of home routines on school routines. Speaking about these possibilities help to explain his behavior in order to inform the next steps in her pedagogical approach.

The relationship and trust I continue to build with this teacher offers the possibility to witness the complexity of issues that may emerge in a language immersion context:

[Karen] asked me what I would do to help him remain on his cot during naptime. We discussed some potential options: Have him engage in a quiet activity like look at a book or work on a puzzle at the table. Another option would be to help him lie down to relax; rubbing his back and staying with him at the beginning minutes of naptime a few times a week could pay off in the long run. Or sit with him speaking in Cherokee or singing a song to calm him down. If she can continue to remind herself that he is three years old and is new to the immersion setting she may be better prepared to find many new ways to connect with him. (Yazzie-Mintz, Field Notes, October 9, 2007)

Karen's willingness to share her frustration and to ask for my thoughts about what she could try revealed a developing trust and relationship that could eventually shape her instructional approach. Having me observe her work during a tough time was not easy. She showed strength even during a time she might have felt she was exposing weakness.

Recognizing weaknesses and transformation can illuminate strength, even though the practice or intent may not be perfected. Examining the complex truths evident in the daily routines of Native teachers' work allows for examination of mistakes. The teachers at the Cherokee Immersion Program certainly reveal the complexity in both defining unique and effective practice in language instruction and identifying issues with which they require assistance, such as increasing knowledge about how to manage a classroom or assist a child in developmentally appropriate ways.

As the Portraitist, I am fully present in this research setting, I participate in verbal exchanges with the teacher as the teacher invites me to engage in a potential instructional response. I empathize with her efforts, sense her frustration, and take note of the impact that my presence has on her pedagogy and engagement with the child. I attempt to develop an understanding of her perspective, as it is the job of the Portraitist to "imaginatively put herself in the actor's place and witness his perspective, his ideas, his emotions, his fears, his pain" (Lawrence-Lightfoot, 1997, 146), an important practice in developing symmetry, reciprocity, and negotiating boundaries between researcher and teacher.

Lawrence-Lightfoot refers to these moments as "unexpected intimacy," a closeness neither researcher nor subject anticipated. In this moment, the process of building relationship—a central process within Portraiture—includes mapping the boundaries between Portraitist and actor, "distance and intimacy, acceptance and skepticism, receptivity and challenge, and silence and talk" (Lawrence-Lightfoot, 1997, 158). The Portraitist's challenge to document what is good, working, and valued is a "generous stance" that "opens up a space for the expression of the weakness, imperfection, and vulnerability that inevitably compromise the goodness" (Lawrence-Lightfoot, 1997, 158). In the previous example, I am not merely interested in "idealized portrayals" of teaching, but rather committed to documenting the tensions, distortions, and supports that ultimately allow me to capture a more complex "expression of strength" (Lawrence-Lightfoot, 1997, 159). The teacher is portrayed as a strong teacher not because she always gets it right, but because she struggles with frustration and with her own capacity to sustain her instructional effort in light of a child's perseverance to avoid naptime.

Cultivating Agency

In my research, I want Native teachers and communities to have agency in the research process. In this study, sharing an educational agenda is the common ground we stand on in my work as researcher and their work as teachers. To treat their work and knowledge as if it were merely only about teaching, without acknowledging the historical contribution, would be less than satisfactory. The language immersion teachers know the scale of their work and seek to complicate their work. The work of the researcher is shaped by the priorities of the Nation and their teachers, while simultaneously mapping this work in the larger inquiry focused on language revitalization. By drawing upon opportunities to create a series of cyclical investigations that will inform instruction and will also reveal new questions, we engage in circles of inquiry. Circles of inquiry lead to deeper and detailed inquiry, with potential for cross-cultural contexts (Yazzie-Mintz, 2006, 3).

Taking tribal nation and teachers' questions and investments in research seriously places them as central actors in the research process. The collaboration seeks taking action on our inquiry within our own and shared contexts. Portraitists "see relationships as more than vehicles for data gathering, more than points of access. They see them as central to the empirical, ethical, and humanistic dimensions of research design, as evolving and changing processes of human encounter" (Lawrence-Lightfoot, 1997, 138).

Defining Purposeful Inquiries beyond the Academy

I presented my research project in a Navajo agency meeting for approval by elected Navajo leaders, providing an example of how tribal people and community members can enact the purpose of research beyond the academy:

> A Navajo woman wearing a teal green velvet blouse, turquoise and silver jewelry draping her neck, her hair pulled into a tightly tied bun sitting in the front row with her sister "grandmothers" stated, "I read your description of the research you are proposing to work with teachers who teach Indian children. In the fourth paragraph you say you will interview teachers to learn about the depths of cultural knowledge that they hold.

What about interviewing community members, parents, and children for the depths of their cultural knowledge?".... While she spoke I heard the urgency to connect inquiry within the classroom with knowledge held by the community. I heard her also stating that cultural conceptions are developed not only in schools but also in homes, within community, with families, relatives, and in realms of leadership. That in these locations, culture is defined by the actions people take and *that* (the actions that people take) is what should be captured in the study. (Yazzie-Mintz, 2006, 7–11)

Seeing and hearing the complexities of cultural research unfold before me, I needed to understand teachers' work, their conceptions of culture within the realm of the classroom than to see and document the many intersecting circles of inquiry *with* community, parents, and children. I could not by myself make the inquiry about culture successful without the intimate perspectives of local community (Yazzie-Mintz, 2006).

Having access to intimate knowledge, pursuits, and political agenda voiced by Native leaders, community members, and teachers allows the Portraitist to expose veiled discussions of education and research in unexpected realms, as within communities. The nature of Portraiture research inquiry is strengthened by their central questions, insights, and resulting plans to act on research findings beyond mere discourse within the academy.

Conclusion

Within the field and practice of American Indian education, Native communities have not held power to decide what education (schooling) should be or how it should be implemented (Lomawaima, 2000). The creation of purposeful research in American Indian education is powerfully shaped by historical and existing Westernized research practices. It seems that progress to produce purposeful research projects that might reveal important findings is stunted by the internal/external debates of insider/outsider perspectives, Native/non-Native location, and/or debates about traditional/alternative research methodologies. These existing tensions work to dichotomize rather than inspire groundbreaking possibilities to imagine scholarly actions that move us all forward to respond to the critical needs of communities in immediate ways. Native researchers whose work focuses on Native education within local contexts have cultivated intimate biases, shifting and altering the balance of power with Native communities.

Intimate biases equip Native teachers to respond appropriately within their respective cultural, social, political, and instructional contexts.

In addition, understanding data within context is an important aspect of interpretation and meaning making by researchers. The holding of biases enhances understanding of knowledge within context, and is a positive attribute rather than one that hinders perspective. Having intimate bias predisposes us to seek and draw upon the strengths of our subjectivity and have a close examination of knowledge and experience acknowledging informed and purposeful perspective.

An example from my own instructional context demonstrates why research with Native communities needs to speak to multiple audiences. Each semester in my role as a pre-service educator at my current institution, I have witnessed a diversity of opinions and questions clearly illustrating that our future teachers have little to no knowledge of the continued existence of Native peoples within the United States, not to mention Indigenous peoples in a broader, global context. I introduce myself to my students as a member of the Navajo Nation, and show how my work within education is strongly shaped by my identity, perspective, and experience. During a routine class exercise, a group of pre-serve teachers worked on drafting ideas for a culturally appropriate history lesson. The students thought that during the month of November they could engage elementary children in an investigation about the first thanksgiving (there is a great deal wrong with this idea), and while they debated and discussed possible ideas, one student asked the rest of the group, "Are Native Americans allowed off their reservations?" I was nearby listening to their developing conversation, when one of the students turned to ask me, "Is it illegal for Native Americans to leave their reservations?" I responded, "I hope not, or I'd be illegal." Their reactions were of surprise, and one indicated, "I did not know Indians still existed."

This story is very important to my inquiry in Native education. While I seek to understand educational structures and pedagogy that support Native existence, identity, survival, and resistance, I also seek to educate dominant society about ways to think about education outside of the structures and experience of dominant society. There is a challenge to helping educators connect purposeful research inquiry with purposeful instructional practice. Employing Portraiture as a research methodology creates opportunity for building relationships among those researched (teachers and Native nations), those who use the research to inform teacher practice (pre-service teachers), and those who conduct research with communities (researchers).

To engage in research for the primary purpose of recovering and acknowledging Indigenous knowledge requires one to accept the possibility of being placed on the margins of the historically non-Native academy by our colleagues, for, without a doubt, "the academy has not historically valued or respected our knowledge" (Wilson, 2004, 73). Current research questions generated by Native and non-Native scholars interested in American Indian/Alaskan Native/Native Hawaiian issues can connect with a variety of interdisciplinary and intellectual interests among tribal nations and within the academy. While we must hold steadfast to our priorities to tribal nations, we must also seek the ways in which our scholarship informs larger methodological questions and has implications for generalized knowledge. Within the field of American Indian/Alaskan Native education, we see that educational inquiry and research methodology are formed and shaped by purposeful interdisciplinary questions focused on Native existence, identity, survival, and resistance.

CHAPTER 11

Lesbian Youth and the "Not Girl" Gender: Explorations of Adolescent Lesbian Lives through Critical Life Story Research

Elizabethe C. Payne

"Girls are made of sugar and spice and everything nice. Boys are made of sticks and snails and puppy dog tails."

Gender and sexuality are not only intertwined, but, as constructs, each is dependent on the other for its loaded heterosexual meaning (Ingraham, 1996; Williams and Stein, 2002; Thorne and Luria, 2002). Resistance to gender enculturation is resistance to the cultural system that pairs anatomy with prescribed gender roles and sexual scripts. A key component in gender resistance is agency. Giddens (1979) states that agency "does not refer to a series of discrete acts combined together, but to a continuous flow of conduct" (55), which is intrinsically related to power and contains within it the recognition of choice—the actor has chosen to act in a certain way but could have acted otherwise. It is a striving for self-determination and mastery of possibility. Agency is "transformative capacity" (Giddens, 1981, 53).

The role of agency in the process of gender construction has received some attention in research (Thorne and Luria, 2002; Carr, 1998). "In these approaches, agents are understood to actively construct gender within the limits of existing social discourses or historically specific social institutions" (Carr, 1998, 528). Drawing upon data from a critical life story study, I argue that lesbian adolescents become aware of their childhood "difference" from same sex peers—and their move to locate themselves outside of the existing gendered options—through the themes of agency and gender resistance. The young women in this study transgress the binary gender boundaries by implicitly creating a third option—"Not Girl."

Previous studies have shown that strict social rules governing gender behavior, such as children's gender-typing play as only appropriate for one particular sex (Greendorfer, 1991), are enforced through childhood peer groups (Grant, 1993; Thorne and Luria, 2002). In middle childhood, the social world is largely segregated by gender. Girls and boys are separated spatially, with boys controlling the large territories of playing fields and girls the smaller concrete spaces for jump-rope and hopscotch. Boundaries between girls and boys are often ritually marked as members of each group identify themselves in opposition to the other group—they are on different "teams" or "sides," even when not formally engaged in competition. In these separate bounded spaces, different subcultures are created and sustained, each with its own rules of relations and patterns of talk. In this way, boys and girls can be said to "occupy separate worlds" (Thorne and Luria, 2002, 129).

Carr (1998) asserts that these rigid forms of gender can actively be resisted and that "tomboys" engage in this resistance. Her life history study with "tomboys" examined the motivations for her participants (predominantly lesbian or bisexual women, of ages twenty-five to forty, recalling childhood experiences) to actively choose "tomboy paths." These included an "aversion to feminine activities and a preference for masculine ones," a dislike of feminine roles and an "awareness of the advantages of masculinity" (535). Carr suggested that through the conflation of gender and sexual identification, tomboy teens may be "nudged" toward lesbian identification (531). Plummer (1996), too, suggests that a childhood sense of gendered difference could later provide "retrospective clues" (71) for an interpretation of self as homosexual, utilizing the dominant cultural understandings of the relationship between normative gendered behavior and heterosexuality. In this chapter I argue that, for these young women, the gendering of self as "Not Girl" in childhood created a sense of self as gendered "Other" that could later, in adolescence, be forced into articulation as sexual difference through the gender and sexual binaries.

Gender identity is tightly connected to concepts of sexuality. In a heteronormative, heterosexist society, the construction of the gender role reflects the assumption of eventual heterosexual pairing (Ingraham, 1996). In this "depiction of reality . . . heterosexuality circulates as taken for granted, naturally occurring, and unquestioned, while gender is understood as socially constructed and central to the organization of everyday life" (169). This heterosexist assumption underlies gender enculturation for both sexes and limits the range of expression and expertise a person of either gender "should" acquire. "Dominant conceptions of sexuality imagine sexuality in terms of binaries: male versus female, heterosexual versus homosexual. Deviations from 'proper' gender are sexualized" (Williams and Stein, 2002, 61). How people perceive their sexuality is connected with their understandings of the self as masculine or feminine (Williams and Stein, 2002). An understanding of oneself as gendered is thus implicated in the understanding of oneself in terms of sexuality and the identity issues linked with it.

Though discussions of agency are increasing in the examination of gender construction, "agentic approaches have rarely been applied to questions of gender *identification*" (emphasis in the original. Carr, 1998, 528). Anthropologist Esther Newton (2000) recounts her "lonely" childhood experiences as troubled by her being "stuck in the girl gender, which is linked worldwide to hard work, low pay, and disrespect." Later, when she identified herself as "gay," she says she was given a second gender, the "gay gender, butch." The young women in this study pose the possibility of lesbian as an alternate agentic gender identity, beyond the boundaries of "butch"—a construction of an agentic female gender identity that began in childhood with an understanding of self as "different" from other girls and an identification of "Not Girl."

In their talk of childhood, the young women in this study consistently contrasted the play activities of girls with those of boys, distancing themselves from "the girls" and placing themselves in gendered nominal limbo through this difference. "Girls do X. *I* don't do that." Through these repeated distinctions, they construct a self as "Not Girl." Though they recalled preferring games considered "boy games" and disliking "girl games," the distinctions they make rest not just upon preferences but upon the gendered characteristics of those who prefer certain games and activities. The differences they mark between themselves and girl peers revolve around a passive versus active, reproductive versus productive dichotomy. This mirrored the girl game/boy game dichotomy they used to construct their "girl/not girl" categories in their stories of childhood. All of the young women in this research recognized their preference for active play and attraction to "boy stuff" as placing them outside the expectations of traditional femininity and marking them as different from "*the* girls." These young women constructed the category "girl" as non-agentic and articulated their difference through a claim to agency. Girls were "scared of getting dirty...scared of doing anything," so they "sat on the side lines...just watching" what was going on (Linda). The distinction between preferred activities of these young women and what "girls" did is constructed through their talk not only as a distinction between an active or a passive position—actively engaging the world, or sitting and watching—but also as a difference in essence of being—"girl/not girl." Thus, as the young women in this research affirm the binary through their understandings of femininity and masculinity, they challenge the binary through their construction of a third gendered option—an agentic female gender, the category of "Not Girl."

Critical Life Story Methodology

The majority of research on nonheterosexual identities is still conducted quantitatively, reducing the self-labeling experience to a series of steps or stages within a model (McCarn and Fassinger, 1996). The limited qualitative research conducted in this area is often marked by interview protocols that focus exclusively or primarily on sexuality. A critical life story approach offers something new.

A life story is a constantly changing oral form of self-expression that requires an audience (Linde, 1993). The story, as told, conforms to cultural conventions of storytelling, thus the types of elements considered worthy of inclusion are culturally bound. In order to make the story tell-able, an entire life must be reduced to elements deemed socially significant enough to explain particular outcomes within the individual's life. The life story is not only the way in which we position ourselves for the understanding and validation of others, but it is also the way in which we examine our own understanding of our past, present, and possible future. The *truth* of the story is in the meaning it holds for the individual who tells the story, the navigation through social situations made possible by the story, and in the implicit cultural structures used to tell the story.

Critical research assumes that "contemporary societies have systemic inequalities complexly maintained and reproduced by culture," that the social structures maintaining inequity are "real" (Carspecken, 1996), and that such systems should be explored in an effort to reduce oppression and inequity. The life stories shared by the young women in this study reveal the cultural schema they utilized to understand themselves as young women in a culture where their identities as female, lesbian, and as youths are not valued positions. "The precise nature of oppression is an empirical question and not a given belief" (Carspecken, 1996, 8). It is the goal of critical research to make "visible" the subtleties of that oppression.

Integrating the life story method of Linde (1993) into the critical methodology of Carspecken (1996) provides theoretical and practical support for conducting life history research within a critical framework. Critical Life Story method differs from other life history approaches in several significant ways: it begins with an emic approach, utilizes Linde's concepts of multiple identity stories, poses few direct interview questions, and uses a critical qualitative analysis. By using only a few interview questions and relying heavily on probes, participants are encouraged to direct the interview to the stories they consider most important, rather than having the focus of the researcher superimposed onto the telling of the story. This differs from many qualitative approaches related to interviews on sexual identity in that there are no questions about sexuality or sexual identity introduced by the researcher. The analysis differs from narrative approaches frequently used in various life history methods in its critical approach to the data and the systems analysis made possible by Carspecken's (1996) methodology.

Participants

Participants for this study were eight white, middle class, late adolescent lesbians, ages eighteen to twenty-one, who had attended high schools in the area of Houston, Texas. At the time of the research, two participants were in high school, one had just graduated from high school, and five were attending college. Participants responded to fliers posted at area college campuses and a community LGBT youth group. The young women had self-labeled them as lesbian between the ages of thirteen and fifteen, with all identifying so by the tenth grade, and all but one doing so without any experience of same-sex attraction or romantic involvement. Though all of the young women who participated in the research claimed a lesbian identity, it is not assumed that all attached the same meaning to the label "lesbian," that "lesbian" is a stable category, or that it has ever had a singular meaning.

Interviews

The critical life story interviews began with a Spradley (1979) "grand tour" style question focused on the high school experience that offered the participants a "place" to start in the telling of their stories. The data desired were not expected to come directly from the answered question, but rather from the flow elicited by the question and from the underlying or implicit assumptions that structured the sense of the stories (Carspecken, 1996). Each young woman was interviewed at least once, with the interviews lasting an average of 3.25 hours. The interviews were audio-taped and transcribed verbatim with the addition of field notes.

Analysis

Each interview was read independently of the other interviews, allowing themes within each one to emerge (Carspecken, 1996). Initial choices about which themes to fully analyze and deem "major" were made based upon frequency, emphasis, perceived emotional response, and connection. Higher level codes were necessary for examining heteronormativity, heterogenders, and the heterosexual matrix. Higher level internally derived codes rely on primary reconstructions to articulate broader cultural themes, relying on the valid mapping of codes to the hermeneutic findings (Carspecken, 1996). This higher level analysis revealed broad social power issues, such as patriarchy, through the higher level codes such as "agency," which the girls genderized through their talk about boys' activities and social hierarchies.

Connecting these etic categories to the data, while remaining faithful to the data itself, occurred primarily through additional exploration of the participant talk within a number of emergent themes, only a few of which will be discussed in this chapter. Peer debriefers (Carspecken, 1996) were used throughout the data collection and analysis processes.

Not Girl: Agency and Active Play in Childhood

The young women in this study recalled their playmates being mostly boys at some point in their youth, but it was not friendships with boys that they emphasized in their stories but boy play, activities, and toys. It was boy "stuff" that they emphasized, and the action characteristics that accompanied that "stuff." They consistently contrasted the play/activities of girls with the play of boys, noting cultural expectations and marking their own sense of identity difference through these distinctions. The stories shared here demonstrate their construction of self as different from other girls through the desire for agency, which all the young women gendered as the absence of what it meant to be "girl." Space limitations allow for the sharing of only a few of these stories.

Amy

Amy remembers herself as a "tomboy" and positions herself as different from other girls, beginning in childhood, recalling that she played a lot "on (her) own," not liking to do "what the girls did."

> I was very tomboyish. I didn't do, like, a lot of what the girls did, like play house and stuff like that. I mean, sometimes, I played with my sister a lot and that was—and she would always play house, or school, or something like that. I would play just because she was my sister and you know, I liked hanging out with my sister, so I'd do that.

Girls played house and school. Amy did not like to do as the girls did. She remembers her favorite childhood game, a detective game in which she would solve mysteries.

> I was a very imaginative child and I would, um, like a lot of stuff I did, I did on my own. (I) pretended that—I would just like play games with myself in my head, like, I was a make-believe character I was—I still even remember the name—I was Jack Wyler and a great FBI agent and I had a partner and, actually, it just comes to mind, while playing this game, I actually had kind of a conflict with myself because, you know, as T.V. goes and stuff, you know, there's always a relationship, you know. So, then, here in my head I'm a male character and I would pretend like I would have, like, a girlfriend, and that would really confuse me. When I was little, I was going, "I'm a girl so I would be going with a guy." So in my head, I'm a guy so I would be going with a girlfriend. So that was kind of weird when I was little.

The strength of the gender binary and cultural emphasis on heterosexual romance can be seen through Amy's childhood confusion over her make-believe character Jack Wyler and his girlfriend. For Amy to be her own TV FBI agent, her character must be male and "as TV goes and stuff, there's always a girlfriend." Here Amy's desire for agency and the gendering of agency as male begins to conflict with the interdependent gender and sexual binaries in such a way as to raise Amy's awareness. To be the action hero she imagined, she must be gendered as other than "girl," which then presents, for her, the issues of attraction and the gender of potential partners.

Amy also called herself a "Lego *person*," stressing the extent to which she identified with and enjoyed play activities she genders as male and claiming an identity position in relation to the activity. "I was a Lego person," she says "I had fun with my brother's toys." She was a "person" who liked active/constructive play. Again, she contrasted her active play with that of the "girls:"

> When I was little, I played mostly with boys, just like robber games. Like we'd have these fake guns, or whatever, and I was, like, one of the few girls that ever played that and I was always playing with the guys. And that was something I really liked to do and really got into and didn't care what anyone thought.

Her statement that she "didn't care what anyone thought" implies that her active play could generate a negative response from others because she was a girl, but that she was willing to risk social sanction for the fun of active play. Later she says:

> I was mainly hanging out with the guys, because I would hang out with the girls, but it wouldn't be fun for me. Like they played Barbies or something like that. That would bore me. And so I played with guys mostly.

She says "I didn't do much with girls." Amy often marks other girls as "they" in her stories, distancing herself from the girl category as if to say: "They are different from me. They are not fun. They bore me. Therefore, I didn't do much with them. I do not count myself among them."

Adults in Amy's life tried to get her to play with "girl toys." Amy says, "Family members—like aunts or uncles—would like—they'd give me a purse or they would give me, like, Barbies and stuff like that, and I wasn't interested in that. I, like, took them apart and used them for different things." Amy tries to make useful what she has no use for, taking apart the Barbies to transform them into something new—turning Barbie into Legos. Amy describes herself not as a girl who prefers different toys, but as a different kind of person—a differently gendered person who acts upon the world around her. She is a female person who transgresses the culturally sanctioned possibilities for girlhood for her own pleasure and power. She is "Not Girl."

Linda

Linda talked a great deal about childhood gendered play and her preference for boy games. With the boys in the neighborhood, she raced "matchbox cars," played "Cowboys and Indians, or" she says, "maybe just cowboys, because I don't think any of us were Indians. We were just running around with these little toy guns." "I always wanted to be the sheriff." She was given a tool set by a neighbor when she was seven and enjoyed taking things apart with it, including one attempt to disassemble the family van. Linda enjoyed acting upon the environment around her. She says, "I have always hung out with guys; I had always been friends with guys. I hated girls when I was little—they were so dumb."

> The girls seemed petty and dumb. I liked to play at recess. I'd play with anybody, but I liked to play with boys more. I liked to use my imagination and the girls were bent on playing, like, they played dolls and stuff like that, but I don't know, it just didn't seem very much fun, so I played with the boys. I played on the swings with the boys and ran around and played leap frog and stuff like that.
>
> All the girls were so scared of getting dirty and they were so so scared of doing really anything, so they ended up sitting on the side lines and just sort of watching what was going on. They were very passive and I just didn't want to be.

Linda discusses and identifies the difference between girl play and the types of boy-identified activities she enjoyed as the difference between active and passive play, between engaging the environment and being afraid. "All" the girls were "scared," they "watched" what went on, "sitting on the side lines." They were "passive." Linda didn't want to "be" those things. She wanted to "be" something other than what "all" the girls were. Linda associated active, agentic play with "being" male, and being "passive" and "scared" with "being" female. Activities are repeatedly constructed as gendered through the dominant discourse. "Leap Frog" becomes male. "Sitting" and "watching" becomes female. Hence preferences for being "active" push toward an identity problem through the gendering of "active" and "passive" as modes of interaction. Difference is experienced not through simple association with "active" or "passive," but association with what is "male" and what is "female," creating a possibility to be outside of categorical expectation through interaction as "active," agentic female. Linda tied "being" to "doing" directly. It was her preferences in the realm of "doing" that marked off different modes of "being" for her. Linda actively chooses to "be" something other than what all the girls "are." She is "Not Girl."

CiCi

As a child, CiCi liked to skate board and go out into the woods. She liked cops and robbers, mountain biking, football, and played with boys or by herself. She did not like to play with girls. "Girls were no fun." Her favorite toy as a child was a "Rambo set and it was like this big knife—big knife and, you know, the head band, and so I'd run around swinging it." She was also fond of her "Karate Kid suit" and her "Knight Rider tricycle." She says "I just needed the outfit and I was set." CiCi's "outfit" transformed her into an action hero.

CiCi's childhood contrast between herself and "the girls" is centered on the types of toys that boys played with, and the action that followed those toys. "Girls just, like, sat around and, like, I don't know, didn't do anything." Later she says, "I didn't wanna do like the girls did, like paint my toenails pink, or whatever, and brush dolls' hair."

She describes the experience of toy shopping:

> There was like the Barbie aisle and there was the boy's aisle. I looked at the Barbies and they did nothing. They're all pink and they're in boxes, and you dressed them up. They didn't do anything, you know? You couldn't run around with them or, like, shoot them or anything. They weren't fun. But you go in the boys' aisle, there's boomerangs, there's guns, there's just, like, balls and everything!

CiCi says she "just loved being outside and running around, and those (boys' toys) were more functional toys." "You actually *do* something (emphasis hers)." CiCi refers to the toy store aisle displaying girls' toys as the "Barbie" aisle, equating "Barbie" and "girl." She remarks on the lack of variety and the inactivity of the toys there ("They didn't *do* anything"), and contrasts it to the variety and the active possibilities of "boys'" toys. She stresses the endless opportunity for active play through the exclamation of "and everything" in reference to the toys present on the "boys' aisle." "Doing" for CiCi is male. Girls' toys, like girls, didn't "*do*" anything. CiCi wanted to "do" something. CiCi is "Not Girl."

Summary

Amy, Linda, and CiCi played by themselves and with boys in activities traditionally marked as belonging to the "boy" domain. They pretended to be characters who were male and who were active problem solvers—good guys, ridding the world, in one sense or another, of evil. They were agents of change. These young women associated themselves with boy activities—not with boys—and they did not express a desire to be boys, but through their attractions to "boy things" they mark themselves as different kinds of people. Physical activity, problem solving, and use of external spaces are here being associated with maleness—acting upon the world is male. Passivity, playing with dolls, fear of getting dirty, sitting on the side lines, and watching as opposed to doing are all associated with girls. Rarely did any of the participants use the phrase "the other girls," as in "the other girls played with dolls." They did not identify themselves as a part of that group they labeled "girls." Rather, they repeatedly reinforce the binary, lining up what "the boys" do and what "the girls" do, making the tacit argument that they do not belong in these gender categories, eventually making a place for themselves outside of one binary and inside another through a desire for an agentic female gender. They are "Not Girl." These young women, as children, were focused on activities, personal preferences, a sense of difference, and a quest for action—not gender identities—but their choices of activity and modes of interaction immediately introduced implicit gender identity issues that would become explicit later in life as they sought to make "sense" of their "difference" in adolescence when gender becomes foregrounded in identity formation.

Gender Orientation and the Implicit "Not Girl"

As the young women in this research outlined what "the girls did" as a contrast to their own interests—their activities emerged as those of an implicit "Not Girl." "Girl" is a semantic category named by the participants and used as a contrasting construct for their descriptions of self. Semantic categories are meaningful structurally as they take their meaning through their relations to other categories within the culture through similarities, binary oppositions, hierarchical subordinations, and so on, with those other categories. As they constructed the category "girl," they implicitly constructed the category "Not Girl," a category with no semantic cultural equivalent. The content of the "Not Girl" category includes resistance to heterogendered expectations of appropriate "girl" behavior, a preference for active play, problem solving, self-determination, fun, and agency. Included also in the category is a claim to broader possibilities of being than those afforded to the category "girl."

The heterogendered structure (Ingraham, 1996) of American culture confronts young people in childhood with the boundaries associated with proper "girl," or "boy," producing a sense of difference for some in early life that could later become forced toward an articulation through existing gender and sexual identity categories. In a binary gender system, the participants' feeling of "difference" has no official place of "otherness." Within compulsory heterosexuality, there is an "Other." So, I am here proposing the possibility that the refusal of heterogendered normative expectations and the quest for an agentic positioning of self, key in the construction of difference in these young women's stories and in their "girl/not girl" categorizations, contributed to an understanding of self as outside the gendered binary. Likewise, the self is located, in adolescence, as a different kind of woman—that their lesbian identification in early adolescence was influenced by this childhood process of self-othering. Self-labeling as lesbian in adolescence is related to the categories that these young women utilize to make sense of their difference. The difference, I believe, begins in reaction to various areas of the "heterosexual matrix" (Butler, 1990) that supports the gender binary, and where the association between their sense of marginality and sexual difference is rooted.

Judith Butler (1990) describes the "unwritten and written codes of heterosexualized gender systems" using the term "heterosexual matrix" to "characterize a hegemonic discursive/epistemic model of gender intelligibility that assumes that for bodies to make sense there must be a stable sex expressed through a stable gender that is oppositionally and hierarchically defined through the compulsory practice of heterosexuality" (151). These "unwritten and written codes" to which Butler refers serve to protect both the gender and sexual binaries from critique.

Ingraham (1996) addresses this heteronormativity and calls into question the reification of sex, sexuality, and gender through her concept of heterogenders. "By shifting the focus from gender to heterogender as the primary unit of analysis, institutionalized heterosexuality becomes visible as central to the organization and construction of gender" (187), and it opens to critique the role of patriarchy in preserving institutionalized heterosexuality. The gender and sexual binaries are not independent of each other, but rather each is necessary to continue the fiction of the "natural" or "normal" in the other.

In adolescence, these young women framed their gendered difference, initially experienced in childhood, through the sexual binary for lack of option within the binary, where the difference was experienced, and constructed "lesbian" as a gendered, rather than an explicitly sexual, identity. The story of this gendering self as lesbian began with the construction of an implicit "Not Girl" self in childhood. This "Not Girl" self reflects their desire to locate themselves as a different type of female, one not subject to the perceived constraints and lack of agency associated with heterogendered expectation. In adolescence, the implicit "Not Girl" semantic category of childhood finds its cultural label outside of the gender binary and inside the sexual binary. As they continue to construct their "Not Girl" category in adolescence, "Not Girl" is, perhaps, named "lesbian" by these young women. Lesbian as a cultural category depends upon the contrast set of other categories for its meaning; thus, "girl" is constructed as heterosexual and "lesbian" serves as gender. "Lesbian" becomes, for these young women, the agentic female place.

CHAPTER 12

Breaking the Hymen and Reclaiming the "Cherry": Adolescent Language Use in Negotiations of Autonomy in a Sexuality Education Program

Cheryl A. Hunter

Introduction

In early adolescence, from ages eleven to fourteen, health education often targets the development of health-promoting behaviors based on research that suggests that during this timeframe health risk behaviors are often initiated (Bruess and Greenberg, 2004). Developmental characteristics of early adolescents center upon biological changes (rapid growth and the onset of puberty), psychological changes (increasing independence and the balance of responsibility and privilege), and social changes (negotiating peer and parental influences). Muscari et al. (1997) suggested that it is the magnitude and complexity of these changes that place the adolescent at risk for engaging in health risk behaviors.

This ethnographic study examined a teen pregnancy and disease prevention program for seventh and eighth grade students implemented within three Midwestern middle schools. The overall stated goal of the program was to create peer-mentors to further the scope of the pregnancy and disease prevention program. This chapter examines the communication and interactions between adult facilitators and adolescents and the role of language in adolescent negotiations of power within the program. I present both data and analysis in an alternative format. The adolescent participants were creative in how they challenged program goals and I wanted to depict this creativity in attempting to actively present them in final writing. Likewise, I present interview data interjected within observation data, trying to not only demonstrate how interview data contribute to the analysis of observational data but also show the necessary richness that observational data provides.

Theories of Adolescence

Adolescence is recognized as a highly susceptible time in which peer influence is pervasive (Hill and Holmbeck, 1986). Adolescent peer cultures are active and creative in that they transform elements of the adult world to meet the needs of the peer world, suggesting that in peer culture: (1) Children make persistent attempts to gain control of their lives and (2) they always attempt to share that control with each other (Corsaro, 2005). Examining adolescent peer groups is not only valuable in that it demonstrates how youths participate with their peers to construct unique adolescent peer groups, but also carries forward to reflect how these adolescents make sense of their lives as often marginalized members of adult culture.

Adolescence has also been recognized as a "time" in a youth's life that creates a liminal boundary. Lesko (1996) argued that adolescents are caught in an "age-structuring system that keeps them liminal in their adolescent years" and both "imprisoned in their time," by their chronological age, and imprisoned "out of time," unable to go backward or forward without negative labels being applied (457). For some adolescents, specifically at-risk youths, the institution of schooling can represent a liminal space that is created and maintained by the institution. For example, rural teen mothers described the liminality of mainstream schooling, explaining how school required their attendance based upon their age, which worked against them when they tried to negotiate attendance policies based upon their condition of pregnancy (Hunter, 2007). The liminal space of adolescence is a reflection of how adolescence has been socially structured, and it results in social parameters that youths must navigate in their interactions with adults, peers, and institutions.

Language is an important factor in understanding social processes within youth peer culture. Similar social processes underlie all intergroup communication (Tajfel, 1978). Youths express their group affiliations in the ways they adhere to the language of youth culture (Fortman, 2003). Youths often reject adults who try to use lexical and stylistic features of youth slang (Platt and Weber, 1984). Platt and Weber (1984) contended that attempts at speech convergence may be "disparaged due to interference from unfamiliar cultural or stylistic strategies but also perhaps more significantly, due to out of group membership of those using the language" (109).

One of the most principal determinants of the quality of adolescent communication is the situation or environment in which communication occurs. However, little is actually known about the ways in which context influences communication (Fortman, 2003). Likewise, if attempts at speech convergence between adolescents and adults end with isolation or rejection by either group, communication between these groups stalls. This has important educational ramifications for health programs that attempt to establish relationships with adolescents, and specifically their peer culture, with the overall goal of adolescents adopting adult behaviors.

Method

This research has employed a critical ethnographic framework. Critical ethnography assumes that all knowledge is essentially mediated by historically and socially constituted power relations, and that facts are not disengaged from values (Carspecken, 1996; Kincheloe and McLaren, 1994). Researchers have recognized that what was once considered "the" scientific method is, actually, embedded with assumptions that supposed facts could be separated from values (Carspecken, 1996). As a result, criticalists counterargue that all researchers, regardless of methodological approach, bring value orientations to their research domain. In keeping with this claim, criticalists identify their research value orientations as a means of keeping potential biases in check and as a way of keeping the process transparent and open to dialogue, demonstrating that research value orientations do not "construct" the object of study (1996).

During this study, I participated as a program facilitator for a teen pregnancy and disease prevention program, therefore it was necessary to reflect upon the value orientations I brought to the analysis. I employed in-depth observational field notes, interviews, and focus group interviews, as well as gathering curricular materials over the span of a full school year. Classes met weekly for an hour in each of three middle schools. I used initial observations of the program and adolescent interactions with each other and program staff as a means to begin organizing themes that were used in interviews. I collected observational data for three months prior to scheduling interviews. Observations included traditional observation techniques as well as participant observation, in which I participated within the observational setting as a facilitator of the program. I interviewed individual participants initially to assess what goals the particular participant had for the program and subsequently, upon program completion, to assess if those initial goals were met and how they were met. I also interviewed adolescents in focus groups during the second semester of class.

The research questions centered upon the following: the outcome goals for the different sets of program implementers (administrators, facilitators, and youths); goals that were met or were not met by different implementers; and the ways different implementer groups gauged the final program outcomes.

Research Site and Participants

Appleton is a small town in a Midwestern state with three public middle schools. Appleton Hospital sponsored and supported a teen pregnancy and disease prevention program called Knowing Your Choices (KYC). KYC was a program funded by state grants that "promoted and supported youths in their efforts to think carefully and cautiously about the decisions they make." According to the program mission, the program promotes tolerance and understanding regarding sexuality while training students to serve as role models for healthy behaviors and responsible choices among peers.

KYC was chosen as the research site based upon my interest in sexuality education and my previous research interest in how at-risk teens navigate the institution of schooling. Upon initial inquiry into the program as a research site, Andrea, the program director and nurse, offered me a co-facilitating position with Martha, who was recently hired to begin implementation of the program. Martha, also a registered nurse, was working on an advanced degree in public health and applied to work with the program out of her interest in sexuality education and specifically her interest in adolescent disease prevention.

The participants were seventh and eighth grade students who applied to participate in the optional program. For one hour each week, students were excused from a class to participate in the KYC program. All students who applied were accepted unless they had previously participated in the program. In the application process, students were asked why they wanted to participate in the program. Most students responded that they were generally interested in knowing more about sexuality, and several applicants stated they wanted information that other adults, specifically parents, were not offering. The curriculum included learning reproductive anatomy, reproductive health and sexuality, disease and pregnancy prevention, and positive self-esteem. A total of thirty-five students (eight male and twenty-seven female), out of a total of sixty program participants (fifteen male and forty-five female), chose to participate in the research study. Adult participants included: Andrea, the program director; Martha, the program facilitator; and Don and Allan; the middle school guidance counselors.

After reflecting on field notes gathered during observations, I created my interview protocol based on general questions that followed up on issues adolescents raised during class and informal conversations that related to the overall research questions. I interviewed the participants in small focus group interviews, conducting eight focus group interviews over the course of the last four months of the school year. Focus group interviews are a particular type of interview in which group dynamics facilitate a focus upon the most important topics with a consistent and shared view (Robson, 2002). Recognizing a limitation to focus group interviews, that the results may not represent the wider population, I balanced this data with individual interviews, observations, and triangulation of data for reoccurring themes.

Validation Techniques

I used multiple recording devices (tape recording and note taking), prolonged time in the field, and followed a flexible observation schedule (alternating observation times and days) as validation techniques. I also used low-inference vocabulary and observer comments, speculations about meaning that is clearly distinguished from objective-referenced data (Carspecken, 1999), in the written observational record. To monitor validity of the analysis and interpretations, I used a peer debriefer during the data analysis phase to examine the transcripts and interpretations for potential researcher bias and assess a level of confidence in interpretations. Likewise, I used member checks, the sharing of notes with participants, to ascertain if participants agreed with recorded data (Carspecken, 1999).

I present my data and analysis below in an alternative format. All observation and interview data from all three schools over the course of the study are compiled to represent one "day" in the health education program. The findings discussed in this chapter focus upon the ways youths used language as a means to negotiate their autonomy in the health education program. There are interschool differences, as well as race and class differences, that are not discussed here. The goal of this alternative presentation is to elucidate the complexity of student and adult interactions. The presentation of data begins with observation notes from the first day in the program implementation in the middle schools, when program facilitators are first introduced to students. Interview data are interjected as it informs the analysis of the observation, and it attempts to demonstrate how interview data contribute to the analysis of observational data.

Findings

Howard Middle School, Thursday afternoon: 12:58 p.m. Bell Rings

The door to room 205, an unoccupied science classroom, swung open wide and a slow trickle of twelve and thirteen-year-old boys and girls entered the room. Most students took no notice of me, Martha or Don (the middle school guidance counselor) at the front of the room and continued either extremely quiet or more noticeably loud conversations. Some sat down immediately, others remained standing. The second bell rang.

Don immediately began to speak in a very low monotone voice, which initially could not be heard over the noise of student conversation. He scratched his head and continued to talk while the voices of the students gradually decreased in volume. Don asked the students to give him their attention, repeating this several times. Finally, raising his voice to a normal level, Don asked again for everyone's attention and then dropped his voice slightly. He reminded the students that "even though [they] were accepted into the program, it was still an elective for [them]." He dropped his face to stare at the students over the rim of his glasses and spoke even softer yet firmer when he said, "And I will know (emphasis) if anyone is causing any problems."

The reference that "I will know what you do" that Don made in front of the students and the program facilitators was one of multiple examples when school administrators or school faculty made explicit, in their use of language, that they would have a knowledge of classroom happenings regardless of asserting how exactly this knowledge would be ascertained. Don never spoke to either of the program facilitators or students, when interviewed, in regard to reporting misbehavior or specific classroom activities to him. Likewise, Allan, another school counselor, told facilitators to "just tell'em I know what's going on and that should keep'em in line." Yet Allan never actually requested facilitators to inform him of behavior difficulties, just to tell the students that he would be informed.

This assertion to both students and facilitators from adults in the school was one example of several claims made that implicitly referenced student autonomy in the classroom. Don's claim to "know" demonstrated an assertion that students lack autonomy in the classroom. The explicit assertion of "knowing if anyone caused problems" implicitly referenced a claim that students were not completely free agents to act in the classroom. Ultimately Don claimed that he had complete knowledge of what happened in the classroom in an effort to exert external control over students. It also referenced a false notion that the program facilitators would report behavior problems to him, creating a divide between the adults (Don and facilitators) and the adolescents. This divide could be interpreted as representing a limited sense of autonomy for the students, implying that the adults were working together to assert control over adolescent classroom activities.

KYC had very specific content and session goals outlined in program materials and reiterated in weekly meetings with the program administrator, Andrea. During the first several meetings, Andrea reinforced to the facilitators that class activities and class discussions needed to "stay close" to program materials in order to accomplish each predetermined goal. Divergence from program goals and materials was not encouraged by the program director.

According to the students, following program materials and prescribed activities and goals asserted that the adults (facilitators and administrators) controlled not only what the students were to specifically learn but also what the topics of discussions would cover. Many students in interviews expressed frustration at not having more power in determining course topics. One student commented, "Why can't we have more say in what we talk about, sometimes I don't want to talk about body parts and talk about how to not do stuff my friends want to do. I mean they do nice stuff too, you know." Another student offered a similar comment, "It's always about you guys." When asked to clarify he explained, "Well, its like you have your own things to do and we're just along for the ride and you never ask us what we really want to talk about," and another student interjected, "like you're afraid to ask us, like we might say something you don't want to hear or maybe you're worried that we'll say something you can't answer."

These examples are articulations of student disempowerment, indicating how students regarded having little control over the program implementation. During another focus group interview, one student said that while she liked the program "because [she] can get out of class, talk about different stuff...it was really a bummer that [KYC] didn't want to know more about what [the students] wanted." Another student said:

> There were a couple times when I thought, oh yeah, we'll get to talk about, like something I thought was going to be cool and then we ended up talkin' about kinda the same old stuff. So I would be like, oh here we go again and not really about what it was that I was thinkin'. Then there were those few times when we got to ask questions and those were my favorite because we could ask anything that we (emphasis added) wanted and you had to answer. We could have done that a lot more.

These are examples of student participants expressing how adults usurped their notion of having a level of autonomy to determine program themes or discussions. The comments of the student participant, such as "like you're afraid to ask us, like we might say something you don't want to hear or maybe you're worried that we'll say something you can't answer," represents an aspect of autonomy. This student was suggesting that in letting the adolescents have autonomy in asking questions or determining program elements, the adults would hear something they "don't want to hear," or be asked a question they "can't answer." The student implied that there are topics or language that youths consider appropriate to ask, that adults would not find appropriate and hence do not ask adolescents for their input. The assertion by this youth was that adults are "afraid to ask" because of what the answer might entail. Therefore, this participant suggested that both youths and adults are restricted. Youths are restricted by the adults, while adults are self-restricting. Adults prevent themselves from hearing something they "don't want to hear," in this instance referencing teenage sexuality.

> Martha cleared her throat saying, "The next order of business" would be to establish rules for the group. Martha explained to students that once everyone agreed upon the rules for the program "this becomes a program for [them]." She asked for volunteers to offer some rules that the group would like to adopt and then asked several students to write the rules in their notebook.
> Martha then added one more additional rule. She wrote on the board "No slang." She described to the students that "everyone should use the correct names for body parts and functions." Martha explained that in all the class sessions slang should not be used. "There are correct names for all the body parts and what they do and we will only use those names." She explained that if students did not know the name the facilitators would tell them the correct word. "It's important to use the right

words for things with your friends so they learn what you know and you can correct them when they use the wrong words."

A "No Slang" rule was requested and repeatedly emphasized by Andrea during multiple planning meetings. "You should establish the ground rules by letting the students make up their own rules, but there is one rule (emphasized) that we always need to add. We need to make sure we use correct terminology instead of slang." Andrea emphasized the "importance of using accurate terminology" in meetings and would often ask facilitators to report back if students were following this rule. In an interview Andrea explained, "We like the students to use the correct words and names for body parts, so instead of guys calling it a dick we want them to say it's a penis. So when a student uses slang we need to correct them immediately and tell them one of our rules is that we use the correct names of body parts not slang." Andrea then gave several examples from her nursing experiences to illustrate the importance of correct terminology. Martha agreed with the "no-slang" rule. During an interview she reflected on her belief in the importance of this particular rule. She remarked, "We can't have people running around calling their body parts strange names. What if these kids go to a doctor and he asks them a question and they don't know what he is talking about . . . ultimately we're here to educate them about what they don't know."

During both interviews and program meetings, Andrea and Martha made multiple references to how they viewed the use of medical terminology in sexuality education programs. "I feel very strongly about this," Andrea said. "It is important that students learn to use correct terminology for body parts and body functions." Both Andrea and Martha wanted to establish with the students the strict use of particular scientific/medical language in the program. Throughout the program the scientific language that both Andrea and Martha used came into direct conflict with the language that youths wanted to use.

> The students write the final rule in their notebooks. The next session objective was addressing "sexual myths" and as a facilitator, I began by asking the students to write on a piece of paper one sexual myth or a related question. Once all the papers had been collected Martha chose the first paper, unfolded it, and read out loud, "Can you get pregnant while on the rag?" Martha repeated the question but changed the wording to "on your period" instead of "on the rag." She then asked the students to give their own answer.
>
> "Yeah sure you can" someone answers. "No, you can't get pregnant when you're raggin'" a girl responded. Martha then interjected that this

question was referring to getting pregnant while a girl was on her period (emphasis) and that indeed a girl could get pregnant "while on her period (emphasis)" and then explained how that could happen.

Martha deliberately changed the terminology from "rag" to "period." The use of medical terminology demonstrated knowledge that the adults in this situation have access to, which students may or may not, indicating not only adult authority in this situation through language use but also how a particular language was granted legitimacy and validity. The use of a particular language and the subsequent attempts at enforcement with the students to use specific terminology illustrated an assertion of a power dynamic that was subsequently interpreted by students as invalidating.

> Martha chose another piece of paper unfolded it and reads: "When your cherry pops you bleed a lot." Laughter follows, then Martha reminded the group that one of our "rules is that we are going to use the correct words for all body parts and this question is referring to the hymen." Hands go up and a student asks, "The what?" "Your hymen is the covering within your vagina that can break when you have sex or can break earlier when you insert a tampon," Martha answered. One student asked, "So when your cherry breaks do you bleed a lot?" Another student offers the answer of "my friend says you bleed a lot when your cherry breaks, like when you're on the rag."

The interchange between Martha's use of medical terminology and the student's use of slang was a common occurrence throughout the program. Martha remarked during an interview that she "felt they were just trying to push my buttons... They know not to use slang and they just keep doing it. I guess they know it upsets me." Andrea had similar experiences that related to Martha. Several sessions that Andrea facilitated had similar interchanges between Andrea and the students. In one session, Andrea said, "We won't continue and I will send you back to class if we aren't going to follow our rules. I know you know that you aren't supposed to use slang."

> Another question followed, "So how do you know if your cherry is popped if you use tampons?" "The word is hymen," Martha repeated. "So, how do you know if your hi- your hi-ahhh- you know, your cherry is popped?"

As the students began to speak using slang and Martha replied using correct terms, the students maintained a hesitance to reply using the medical terms. They stumbled over the words and exaggerated them. This hesitance, or resistance, to use medical terminology was a question addressed to students in focus group interviews. Students described how their own use of language was deemed invalid by program facilitators and administrators and subsequently resisted by the students.

All students agreed in one focus group when a student said, "You guys never use our language!" [Researcher: Could you explain what you mean?] "Ahh, you know . . . you guys come and we want you to tell us stuff, all the things that other people won't tell us about, but then we have to sit around and talk all, like, you know, penis this or vagina that . . . I mean who talks that way? My parents (emphasis) don't even talk that way and they won't talk to me about sex either."

Another student followed, "Maybe if we could just be a little more, well, a little more, like's its okay if we don't always have to think about what words we need to use. So we could just call things, you know . . . " [Researcher: Call things?] "Well, call things how we (emphasis) call things." [Researcher: So, do you mean I should use slang?]. A student quickly interjected, "Sure." But then several other students wanted to qualify the assertion of facilitators using slang when they offered the following comments. "No, I don't think that's the best. It's more like wanting you to use slang if you use slang. Like do you talk about it that way?" Another student said, "I guess it's like letting us do what's the best for us. If a person wants to say it that way that's fine but they shouldn't have to say it your way."

In another focus group a similar conversation followed. One student said, "You know what would be great? For you to talk more like us, so you could ask us what we already know." [Researcher: You would like for the facilitators to use slang?] Another student answered, "Well, it's not like we want you trying to sound like us all the time. But at least let us (emphasis) sound like us, you know, you know what I mean." Another student explained, "It's like, I'm not going to ask you some question about a vagina and if you have to ask me questions about oral sex, and I want to say it's a blow-job, that's okay. That's what we mean."

The student's use of their own peer vernacular was a means of resistance toward the adults in the program but was also mentioned as a means to bridge the adult/adolescent culture gap. While the students did not necessarily want the program facilitators to use the peer group vernacular, they also recognized that the adults were limiting student language use and placing limitations on their own autonomy. Students who remarked "let us sound like us" refer to allowing the adolescents the autonomy to decide upon their own language use. The use of language reflected the youths' feelings of wanting to assert their own knowledge and to have that knowledge validated. This implies that youths did not want complete isolation from adults, but rather recognition of a level of autonomy within a mutually respectful relationship.

> "You may never know when your hymen breaks because you may or may not even bleed," Martha answered again. She then pulled one more slip of paper from the pile of questions, opened the paper and hesitated. Her face turned a dark shade of red. She handed the slip of paper to me and said "Maybe you should read this one." I read the slip of paper and laughed out loud while the bell interrupts…

Discussion

The findings suggest that both adults and adolescents use language as a means to negotiate their own autonomy within sexuality education programs. Adolescent language was a means to challenge adult instructors to deliver youth-driven content. Adolescent resistance to content and delivery reflected notions that adolescents wanted more empathy within an autonomous relationship as opposed to a traditional subordinate relationship with adults. Adult participants had very specific goals, based upon scientific scripts, and were less likely to deviate from program goals to meet adolescent needs. Adolescent participants, in wanting validation of their experiences and in asserting their own vernacular, demonstrated how youths resisted program goals and contended they were autonomous agents in the implementation process.

Adolescent language use, specifically the continued use of slang, was a means to challenge adult instructors to deliver youth-driven content. For example, one of the outcome goals for the program administrator and one facilitator was the adoption of "correct terminology," which was not a goal shared by the adolescents in the program. Slang was a means by which youths attempted to appropriate program goals and more closely align them to the adolescent's goal of youth-driven content. In the end, the youths maintained their rejection of "correct terminology" yet failed to achieve their goal of youth-driven content.

Facilitators maintained strict adherence to the "no-slang" rule in course content and progression without incorporating youths' ideas or requests. Program administrators and facilitators failed to recognize their privileging of access to only one acceptable discourse, medical terminology. This finding illustrates what Marshall (1999) articulated in suggesting that privileging only one speakers' truth served to displace all other alternative perspectives.

This study provides significant contributions to understanding effective ways to formulate sexuality education programs that address youths' needs. Sexuality programs that result in adolescent resistance, or end with isolation or rejection of program goals, stall genuine progress for health and sexuality education. This has important educational and policy implications for health and sexuality programs that attempt to establish relationships with adolescents, and specifically their peer culture, and with the overall goal of adolescents adopting adult behaviors.

CHAPTER 13

Disenfranchisement and Power: The Role of Teachers as Mediators in Citizenship Education Policy and Practice

Debora Hinderliter Ortloff

In recent years, globalization processes, in the form of demographic changes, economic migration, the expansion of transnational and supranational allegiances, and (perhaps consequently) increased debates about ethnic and religious rights have presented new challenges to the nation-state. Indeed, the global era has induced states to question, and in some cases redefine, long held notions of citizenship and belonging (Klopp, 2005; Gutmann, 1997). Such notions have traditionally been taken for granted as legislated domains under the control of the national government. Whole groups of people, based on their heritage, immigration status, or other arbitrary variables, have been disenfranchised from participation in the democratic process of their state of residence. Other institutions, such as schools, have taken up the legislated definitions of "who can belong" and used them as the crux of their own policymaking with regard to citizenship education policy. In fact, citizenship education policy-makers have engaged in nationally (if not nationalistically) defined norms of belonging (Keating, Ortloff, and Phillipou, forthcoming) and now are faced with the task of reconsidering these definitions. As Banks (2008) concludes, there is a need to prepare young people for a transformative citizenship that, as Law (2004) notes, requires students to think beyond borders. This necessarily challenges the exclusivity of citizenship as a national domain.

While this pressure to change has not resulted in any country aban-doning local and national citizenship education in favor of a purely cos-mopolitan or global model, globalization has influenced educational reform and discourse (Keating, Ortloff and Phillipou, forthcoming; Sutton, 2005). Indeed, as Banks (2008) demonstrates, one of the most critical areas of influence has been in the emerging recognition that cit-izenship education and multicultural education need to be coupled as areas of educational inquiry policy and practice. He calls for an expan-sion of the traditional citizenship typologies (Marshall, 1999) to include a more transformative multicultural citizenship that recognizes global immigration and migration as part of a new landscape of citizenship. This expansion would provide, according to Banks, educators and schools with a new paradigm for conceiving multicultural and citizen-ship education. Sutton (2005) argues that throughout the world, and in a wide variety of educational systems, multicultural education has become a larger part of the educational reform discourse.

> Loosened from its mooring in the United States civil rights movement, multicultural education has become a rubric—or foil—for a certain arena of educational reform discourse around the world.... [T]he "epochal" dimensions of globalization such as wide-scale human migra-tion and intensification of global communication have complicated social identities within many nations and so stimulated public debate on how pluralism is recognized in the curriculum and pedagogy of national school systems. (2)

Banks' movement toward including citizenship as a part of multicul-turalism is particularly noteworthy because of his preeminence in the field of multicultural education. Previous work has not recognized citi-zenship as a variable in how students are educated multiculturally, but much contemporary work has begun to move in this direction (Banks, 2004, 2008; Walat, 2006; Schissler and Soysal, 2005). Indeed, it is now widely recognized that citizenship, and thereby citizenship education, marginalizes and disenfranchises in a manner that requires active inter-vention on the part of education policy-makers and educators.

While this recognition is critical in understanding how the concept of citizenship has begun to change, and how this needs to be translated into education, it does not provide a theoretical framework for examin-ing how marginalization comes to be reproduced in citizenship educa-tion. This chapter begins to fill this lacuna by offering one theoretical perspective on the production of citizenship education policy as the means of interrogating how marginalization can (and cannot) emerge.

As a part of this larger book, this chapter asks readers to remember that any variable, whether it is ethnicity, religion, skin color, gender, or citizenship, which seeks to disenfranchise and devoice, serves a marginalizing agent. Empirically, it becomes necessary to interrogate not just the existence of the margins, but also the process by which they come into being. Here, I begin this process by explicating an exemplar of the development of a theoretical framework. The theory, and its application to the production of education policy, becomes my data and it seeks to provide a substantive (rather than methodological) means of understanding marginalization through theory. This substantive theory could, however, be used to inform methods.

In short, I will argue that in order to study citizenship education, and take seriously the power of the state to use citizenship education as a means of marginalization, we need to develop new theoretical means of conceptualizing the citizenship education policy production cycle. I will present, in response to this need, a brief interpretation of Habermas' system and lifeworld as a diagnostic tool for citizenship education. I conclude by arguing that studies designed using this conceptualization are able to better probe the complexity of citizenship, diversity, and education, and enhance our understanding of how citizenship education may serve to marginalize and disenfranchise. This exemplar of theory development relative to citizenship education provides a framework that could inform methodological techniques. Such techniques would be more effective at probing citizenship education for its potential to marginalize.

State-sanctioned Marginalization: Who is Allowed to Belong?

Citizenship and citizenship education have the potential to marginalize, in that they institutionalize a form of citizen as best or ideal. Likewise, as is indicated by the use of the word "institutionalize" to describe citizenship education processes, citizenship education, more than other manifestations of education, involves the interaction of multiple levels of state (in Habermasian terms, represented by the "system") with the everyday (Habermas' "lifeworld"). The concept of citizenship is power laden. Nations exercise the power to determine who can or cannot belong through citizenship and immigration policies. Education systems reproduce this concept of belonging through citizenship education policy and practice. As a result, power becomes an inevitable part of this analysis. How power manifests in any given policy production cycle differs on the basis of the national, local, and, arguably, supranational context.

Using Germany, which is where my own empirical work is located, as an example makes this idea clearer. At the national level, citizenship policy in Germany has determined who is allowed to be German and what the notion of Germanness means. Until very recently, German national citizenship policy held that only those with German heritage (so called blood-based citizenship) could be, or could become, citizens. Those born in Germany to non-German parents (or grandparents or great-grandparents) could not be and could not become, German citizens. In turn, this idea of an ontological Germanness (Ortloff and Frey, 2007) has been codified in education policy by the individual German state (e.g., Bavaria or Brandenburg) through required textbooks, curricula, and teacher training. With the advent of the European Union, we see the Supranational imposing a new notion of belonging and Europeanness upon the citizenship policies of its member states and, likewise, the state-level education policy creating a system for cultural reproduction for their definitions of Europeanness and Germanness. These notions result from labeling those that are non-European and non-German as "other," even if that "other" is a member of the European or German territory. To this end these labels negate and exclude, and can be regarded as serving discriminatory practices.

This complexity, although in some ways particular to the German case that I use as an example, is not wholly unique. There are policy and practice tensions between citizenship and immigration (which are usually regulated nationally, and increasingly supranationally), and citizenship education, which is often regulated and certainly interpreted by state and local entities. Caught in all of this, and not to be overlooked, is the fact that young people, in this example ethnic Germans and those of immigrant backgrounds, are the ultimate "beneficiaries" of this policy production cycle. The degree to which citizenship education is able to actually balance diversity and unity in a shifting global landscape has real consequences for these young people. Ultimately though, this multileveled policy production cycle presents substantive questions such as: what should citizenship education look like? what are the conflicts between teacher practice, state curriculum, and federal citizenship mandates? what practices might improve the inclusion of minorities into citizenship education? It also produces theoretical queries such as: how can we conceive of studying citizenship and citizenship education in a manner that conceptualizes these different levels and different powers, while still considering that citizenship is a policy domain that has been and can be used to disempower and disenfranchise? It is this point that I want to explore in a possible theoretical perspective, from which qualitative studies in citizenship education could emerge and have the potential to better understand the impacts upon both the marginalized citizen and how citizenship education marginalizes.

Theoretical Framework

I argue that by explicating the structures, interactions, and concepts within a theoretical framework adopted from Habermas' (1984, 1987) seminal work, the *Theory of Communicative Action*, we are better able to conceive of how citizenship education policy processes work, and this allows us to take more seriously the ability of these interactions to marginalize groups. In this limited space, I have summarized Habermas' major ideas in order to spend more time explicating in what ways we need to augment his work in order to use it to understand citizenship education and marginalization. For a full discussion of Habermas' theory as it applies to this area, see Ortloff (2006). I augment Habermas' work with discussions of symbolic interactionism using primarily the work of Fraser (1989) and authority based on Gutmann (1987) and Weber (1985).[1] As discussed above, I use the case of Germany to develop examples and applications of this idea.[2]

Citizenship education, by design, institutionalizes a particular image of citizen as best or ideal. Drawing on the German case, Habermas (1996) argues that citizenship has been used as a means of both creating and enforcing an ethnocultural community. These communities, as Anderson (1991) points out, exist not through political or economic borders or treaties but through myths and symbols reproduced through institutions and discourses. Thus, in Germany the preservation of the ethnocultural citizenship standards would result in state mechanisms, in particular, education, actively creating and reinforcing myths and symbols of the nation-state (Anderson, 1991; Hobsbawm, 1992). As civil servants, teachers are an extension of the state in this endeavor. However, it would be inaccurate to assume that teachers act, in their interpretation of state policy, without agency (Levinson and Sutton, 2001). Likewise, it would be inappropriate to examine educative processes and their dependence on both state mechanisms and policy as separate from the personal interaction and meaning-making of teachers. Schooling includes and mediates communication and interaction of the state with schools and classrooms.

In Habermasian terms, education is a complex system that links, and indeed mediates, between the system (in the case of my work, the

Bavarian and, by extension, German and European state) and the life-world of families and communities (such as those in the Franconian region of Bavaria). Teachers, among others, are the central cogs in the process. They are the ones who interpret policies, mandates, and goals that manifest in curricular frameworks and directives. They are also the ones who are held accountable for these actions. With regard to citizenship education, it is even more obvious that this mediation between system and lifeworld is necessary and crucial for the survival and future of a democratic state. Habermas' notion of system and lifeworld explicitly includes a private and public dimension. In my adoption of the Habermasian model, this means that my analysis is limited to the public arena of the lifeworld and the public domain of the system because these are where citizenship education is located.

Given this understanding, we can view education within the lifeworld through, for example, the interpersonal interactions of teacher and pupil as contributing to the socialization of a new citizen. Likewise, the system can utilize education as a publicly administered bureaucracy to create citizens for participation in public decision making. Thus, the notion of citizen itself could be both rooted in the lifeworld and in the system—both a reflection of micro-level norms, values, and traditions—while simultaneously linked to system-level constructs such as the economic system in a country (e.g., a citizen has economic privileges that a noncitizen does not).

Habermas interprets the citizen within the public lifeworld-system relations. This study draws from this notion to posit more specifically the interaction between system and lifeworld that must take place in order to produce a citizen. I posit that teachers, as implementers of state policies on citizenship education, are—to a great extent—the producers of citizens. They teach values, traditions, and norms, which are lifeworld notions. These notions, however, attach to the system through system-level processes, such as textbook development, and system-level economic values. Expanding on the latter point, the lifeworld norm of punctuality, which is often held up as being particularly "German," emerges in the economy when punctuality is demanded of workers. Those people who do not represent the norm of punctuality in the standard "German" way do not receive or retain jobs. In short, those who are citizens, both in the legal and normative sense, receive economic benefits; citizenship education is the mechanism by which the normative ways-of-being are conveyed to future citizens. That is, understanding educative processes requires us to consider both system and lifeworld intentions and practices. The public sphere comprises school life and social studies lessons,[3] the latter being given by teachers on the basis of state-level curricula and textbooks. The system is constituted primarily by state-level education policies, because it is these policies that provide direct action imperatives to teachers. Thus, based on Habermas' notions of system and lifeworld, we can examine the production of citizens through the interaction of teachers and the state. In the following section, I will more closely examine this idea of interaction in Habermas' work.

System and Lifeworld: Interactions

The value of Habermas' framework for studying citizenship education is that he emphasizes the reproductive dynamic between the public lifeworld and public system. It is here that I augment Habermas' theory in order to make it more useful for explaining the possibility of marginalization.

Understanding education as a publicly administered system, in which state imperatives are translated into education policy on citizenship education, reveals how the system influences values, such as what it means to be German, "even though those values are themselves negotiated in the lifeworld." This idea of the interface between system and lifeworld harkens to Habermas' (1987) idea of symbolic reproduction. Whereas material reproduction prepares people to be workers and consumers, symbolic reproduction recognizes education's role in preparing students for political participation and opinion formation. Indeed, I argue that education is a manifestation of Habermas' symbolic reproduction. Fraser's (1989) work highlights the distinction between symbolic and material reproduction more distinctly than Habermas' original contributions. For the purpose of this discussion, symbolic reproduction is critical because it guarantees the survival of society and epitomizes the role of the teachers and the place of citizenship education.

Given this concept of symbolic reproduction, it becomes easier to see how, through system-lifeworld interaction, the system expresses influence on value commitments. As Habermas explains, the administrative system, in this case the Ministry of Education, uses—in addition to money in the form of taxes—also the power medium to influence the public sphere, in this case, teachers and students. Ultimately, through system-level political decisions, such as adoption of a limited set of approved textbooks, the passage of new curricula, and the state-controlled training and hiring of teachers, the state can influence the lifeworld-level value commitments of teachers relating to citizenship education. For example, the process would begin through the extension of state power, the state puts forth the image of an ideal citizen through the symbolic reproduction of textbooks, curricula, hiring, and training of teachers. Then, in so doing, the state has the potential to exclude or include possible images of the ideal citizen. Through these political decisions, the system then commands back from the public sphere mass loyalty. As a result, mass loyalty then reattaches lifeworld notions like values and norms to the system. While Habermas is not specifically addressing education or citizenship education, it is easy to compare his conceptualizations of system and lifeworld to the idea of education as the most direct means of the state to influence its future citizens.

Finally, this understanding of the location of education within the system-lifeworld necessitates considerations as to what is meant by positing teachers as mediators. Indeed, it is important to explicate this mediator role of teachers as agentic.[4] By this I mean that teachers as mediators act with agency because of their dual role as civil servants and members of the lifeworld. Teachers, though, have the opportunity to take up, interpret, and reform these system level value commitments within their own lifeworld. Agency is a lifeworld notion that allows teachers to fulfill their role as mediators between system and lifeworld.

To explicate this idea in more detail, it is beneficial to consider Fraser's (1989) critique of Habermas. The critique by Fraser (1989) is premised on the idea that Habermas' framework is useful but has what she calls "blind-spots" (119). She argues that Habermas does not consider gender, and therefore, his critical theory is incomplete. This incompleteness, however, does not nullify the work. As with gender, Habermas' work likewise has blindspots in terms of ethnicity and, as Fraser points out, race. Unequal power relations and the disenfranchisement of non-Germans imply that the state policy on citizenship education has the potential to reinforce an ethnocultural ideology. The power mediums explained above could systematically ignore non-ethnic German residents or, intentionally or unintentionally, reproduce an ethnoculturally gender-biased image of the citizen.

On one hand, teachers who are state servants, and, on the other hand, teachers who are engaged in the personal everyday interactions between pupil and teacher, which characterize the lifeworld, both serve as mediators of these normative claims. In their interpretations of state policy, the teachers can serve as mediators of values for achieving a "communicatively achieved interaction" (Fraser, 1989, 135). That is, the teachers, within the lifeworld, could expand or reject system-forwarded images of the ethnocultural citizen. A multicultural citizen, a cosmopolitan citizen, or a combination of a variety of citizens could emerge as a result of the interpretations the teachers make of state curricula, which ultimately would then result in a different citizenship education from that "intended" by the state. So, in summation, the teachers' role could be a location for examining how marginalization and disenfranchisement may happen or could change.

In Fraser's (1989) interpretation of Habermas, actions that are coordinated by "explicit, reflective, dialogically achieved consensus" (139) are communicatively achieved. However, as the data in my study reveal, the teachers could also reproduce system imperatives rather than interpreting or renegotiating them (Ortloff, 2007). In Fraser's language, then, this reproduction is normatively achieved and communicative action is not acted out. Such normatively achieved action is "tacit, prereflective and pregiven consensus" (Fraser, 1989, 135). The former requires teachers to generate new meanings and a new image of the ideal citizen, either as an extension of the state's image or, in theory, in contrast to it. So the teachers reflect on system-level notions of citizen within their lifeworld and then have the agency to accept, reject, or reinterpret them. This also supports the idea that teachers mediate between system and lifeworld. Empirically speaking, we would expect either patterns demonstrating reproduction or patterns demonstrating resistance to emerge in the teachers' explanations of their interpretations of state citizenship education imperatives. Habermas' theory, despite Fraser's critique, does create the abstract space for emancipatory or communicatively achieved actions. Whether and in what manner this space is taken up by the teachers in concrete actions can be answered through data analysis.

Understanding the limits to teachers' ability to act autonomously is critical to the task of empirically probing to what degree they actually serve as mediators. By using Weber's (1968) ideas of authority, a clearer picture of this tacit agreement between state and teacher emerges. From both the system and the lifeworld, certain constraints are placed on teachers' authority. Weber's notion of bureaucratic authority elucidates the limitations placed on teachers by the system. Weber (1968) writes:

> From a purely technical point of view, a bureaucracy is capable of attaining the highest degree of efficiency, and is in this sense formally the most rational known means of exercising authority over human beings. It is superior to any other form in precision, in stability, in the stringency of its discipline, and in its reliability. It thus makes possible a particularly high degree of calculability of results for the heads of the organization and for those acting in relation to it. (223)

Teachers as employees of the state hierarchy are the extended arm of a bureaucracy and occupy a lower level in the educational hierarchy. Dougherty and Hammack (1990) interpret this idea of bureaucratic authority in school structures: "organizations with bureaucratic authority systems are governed by legal-rational principles that rest on law and on assessments of the 'best' way to delegate responsibility and authority to achieve the required task" (169). Thus, bureaucracies are also supposed to counter unfair practices and are supposed to be neutral.

In the German case,[5] this educational hierarchy begins at the state-level as the Ministry of Education officials interpret federal and EU policies in the creation of curricula and the adoption of mandated textbooks. However, teachers also command professional authority. As Weber (1968) argues, bureaucratic and professional authority are not opposed. Rather, because they both place expertise as central, they are fundamentally complementary. Professional authority, which is granted through extensive state training, is a prerequisite for parents agreeing to entrust their child's moral education to teachers. The principle of academic freedom, which is central to the conception of the teaching profession in Germany (Westbury et al., 2000), supports professional authority of teachers. The existence of both bureaucratic and professional authority underscores the position of teachers as mediators between system and lifeworld. As such, these ideas also reinforce the notion that teachers have the opportunity to symbolically reproduce system imperatives or resist them, thereby renegotiating new conceptualizations of citizenship and diversity. Authority in its various manifestations must be in place for teachers to take up the roles as mediators between system and lifeworld.

Understanding teachers as mediators between system and lifeworld by virtue of their authority leads us again to consider the notion of disenfranchisement and power. Gutmann (1987) puts forth four conceptions of authority over education: "the family state, the state of families, the state of individuals and the democratic state" (19–47). Germany aligns most closely with the fourth conception, the democratic state. In discussing this conception, Gutmann (1987) writes:

> Educational authority must be shared among parents, citizens and professional educators even though such sharing does not guarantee that power will be wedded to knowledge (as in the family state), that parents can successfully pass their prejudices on to their children (as in the state of families), or that education will be neutral among competing conceptions of the good life (as in the state of individuals). (42)

Parents, citizens, and educators are called upon to make educational decisions and not delegate this function exclusively to the state. The education system in Germany, particularly Germany's status as a social democracy that grounds itself in equality, could fail on two of Gutmann's key points. As Gutmann explains, the democratic state does not necessarily ensure equitable education. She posits that the state in which all three stakeholders (parents, teachers, and citizens) democratically choose to exert authority over education by proxy must be nonrepressive and nondiscriminatory. Nonrepressive means the state may not use the education system to reduce or remove choice and agency from individuals or groups. Nondiscrimination requires the state to ensure an equal and adequate education—for all students,—an education that will allow students to participate in the democracy. From a methodological standpoint, we therefore need to examine not only the interactions of system and lifeworld vis-a-vis the mediating teaching but also probe these interactions and interpretations based on Gutmann's standards of nonrepression and nondiscrimination.

Conclusion

This chapter, an exemplar of substantive theory that can inform methods, began by explicating theoretical concepts central to the inquiry. Specifically, this chapter has probed Habermas' notion of system and lifeworld, positing education as an institution that is situated between the system and lifeworld. The dynamic interaction between the two public spheres in Habermas' schematic is critical to understanding the production of citizens, because it is achieved through this interaction. Teachers, via their bureaucratic and professional authority, serve as mediators and make autonomous choices, which either symbolically reproduce state system-level imperatives or revise these commitments within the individual lifeworlds of the teachers.

I argue that this framework can be used to probe both system and lifeworld notions of citizenship. Results can reveal that the nature of citizenship education or the image of an ideal citizen offered to students is constrained by both state and teacher interpretations (Ortloff, 2006). It follows then that the inclusiveness or exclusiveness of this ideal citizen image can best be revealed through a two-tiered analysis that uncovers both the lifeworld and system interests. Such studies need to likewise be reflective of power and oppression, as the reproductive nature of system and lifeworld interactions leave open the possibility of reproducing oppression in the form of discrimination and repression.

Studies that take Banks' (2008) call to consider new paradigms in citizenship and multicultural education need to define a theoretical base that can be used to inform methodological decisions, such as the one I outline here, which allows for consideration of policy production as well as practice. Without taking seriously how power is wielded by the state, studies of citizenship education will fail to really consider marginalization. Consequently, education will be allowed to continue to teach from a perspective that seeks to marginalize those noncitizens in their classrooms.

Notes

1. There are, of course, several key limitations to this analysis. Most notably, I am not seeking to explicate Habermas' *Theory of Communicative Action*. I take up Habermas' social theoretical framework as applied only to the inter-action between the public lifeworld and the public system because, as I argue, this interaction pertains to citizenship education. My application of Habermas' social theoretical framework remains focused on discussing how education and citizenship education can be better understood as an interaction between system and lifeworld.

2. Germany is particularly appropriate as an example case because it has only recently (2000) moved to change its heritage-based citizenship laws. Its membership in the EU also provides a very concrete example of the supra-national level.

3. Of course, the idea of the public sphere in education is much more multilay-ered than this, but because I am at this point stressing the interaction between level policies and teachers' interpretations of them, I have limited my portrayal of the lifeworld to the aspects that are being addressed.

4. Here I am using this word to emphasize the process of enacting agency. Payne (2005) uses agentic in her description of lesbian girls making active choices against stereotypical girl roles. This complements Bourdieu's (1984) ideas of activating cultural capital in which he emphasizes choice.

5. Notions of authority differ from system to system, but Weber's typologies are a useful means of understanding these differences and allowing compar-ative analysis across system lines. Likewise, it allows us to understand how Habermas' ideas of system need to be adjusted based on any given system's peculiarities.

CHAPTER 14

Quantitative Approaches as a Bridge from the Invisible to the Visible: The Case of Basic Education Policy in a Disadvantaged Nation[1]

Aki Yonehara

Introduction: Eyes to the Margins of the World

According to UNICEF (2008), over 77 percent of the world's countries are categorized as underdeveloped. Since the World Educational Conference in Jomtien, Thailand, in 1990, "Education for All" has become a worldwide slogan of educational development; the World Education Forum in Dakar, Senegal, in 2000 adopted "The Dakar Framework for Action Education for All"; the United Nations General Assembly in 2000 adopted "Millennium Development Goals," which counts educational development as one of the central goals. While many educational scholars have contributed to the understanding of international development, these recent frameworks of educational development indicate that educational problems in developing countries have not yet been solved. Rather, the imparity of the educational situation between developed nations and developing nations has increased. This begs the question whether educational research needs new tools of analysis as it seeks to contribute to educational development in the new millennium. In this case marginality is increased not only because the peoples involved are defined as marginalized, but also because theoretical and methodological traditions may have been confined.

Based on this premise, this chapter considers as an example educational development in underdeveloped nations, particularly Tanzania, a sub-Saharan nation, from a perspective of educational *policy*. Policy studies can be an effective approach when examining an educational situation from a macro perspective to identify problems and to solve them. Wildavsky (1979) insists that "policy analysis creates and crafts problems worth solving" (386). According to him, "problem solving for the policy analyst is as much a matter of creating a problem (1) worth solving from a social perspective and (2) capable of being solved with the resources at hand" (388). How can we create "a problem worth solving and capable of being solved?" One of the core functions of educational policy research would be to propose guidance to answer this question. This chapter aims at providing such guidance, through the use of a quantitative approach, using a case study of Tanzania.

From Human Capital Theory to Human Development Theory

Before turning to the quantitative analysis of Tanzania's case, it is necessary to review the theoretical literature related to international development. Meier (2000) categorizes the theoretical works of development economics over the past fifty years into two generations: the first generation (1950–1975) and the second generation (1975–2000). According to Meier (2000), the first generation can be characterized by their "grand models of development strategy" (13–14). Development economists in the first generation took a macro approach toward the issues of international development. Nurkse (1953) describes "the vicious circle of poverty" by saying that "a country is poor because it is poor." Rostow (1960) hypothesizes "the five stages of development," which assumes that a society should follow the hypothetical stages to achieve an economic "takeoff" to "the age of mass consumption." These grand theorists in economics use the term "development" to mean "economic growth." Along with these grand models, it was widely believed in the early 1960s that the benefits of economic development would automatically "trickle down" (Hirschman, 1958) to the field of social development, including education.

On the other hand, the second generation "looked at the growth process in a more microeconomic fashion" (Meier, 2000, 18). The second generation's standpoint can be represented by the approach of human capital theorists. According to Schultz (1971), investment in human capital enhances laborers' knowledge and skill, which contributes to economic growth. He says that human capital is composed of education, health, and other factors that improve the quality of labor. In the 1970s, this theory was incorporated into strategies of international development, constituting the Basic Needs Approach (Streeten, 1981). The Basic Needs Approach is a moralistic rejection of the failure of the macroeconomic growth models created by the first generation of development economics.

After two generations struggled with difficulties of international development, Sen (1997, 1999, 2000, 2002) developed human development theory to provide an alternative perspective on international development. In this theory, he defines the meaning and objectives of development as given below:

> Development can be seen, it is argued here, as a process of expanding the real freedoms that people enjoy. Focusing on human freedoms contrasts with narrower views of development, such as identifying development with the growth of gross national products...Growth of GNP or of individual incomes can, of course, be very important as *means* to expanding the freedoms enjoyed by the members of the society. But freedoms depend also on other determinants, such as social and economic arrangements (for example, facilities for education and health care) as well as political and civil rights (for example, the liberty to participate in public discussion and scrutiny). (Sen, 2000, 3)

Human capital theorists value productivity and efficiency, which are evaluated by the amount of earnings or any factors to promote productivity. On the other hand, Sen evaluates the quality of human life by one's "well-being," in other words, how many actual choices he or she can have for enjoying a better life (Sen, 1999). Sen's question is how to improve the "well-being" of each individual rather than efficiency or productivity. However, it is important to note that human development theory never denies the value of human capital theory (Sen, 1997). Human development theory reminds us of other aspects of human life beyond economics and establishes human well-being as the *end* and economic wealth as a *means*.

When thinking of education as a human right, as an opportunity, and as an entitlement for one's "well-being," literacy can be one of the fundamentals. The association of literacy with health, nutrition, and other social goods has been widely accepted by governments, and "high rates of literacy have taken generations to achieve" (Wagner, 1990, 1992, 21). UNESCO (2005), a leading sponsor of literacy development programs, declared on international literacy day, September 8, 2005, that "literacy is inseparably tied to all aspects of life and livelihood. Literacy is at the heart of learning, the core of Education for All and central to the achievement of the Millennium Development Goals." In this sense, literacy can be considered as a critical part of basic education and thus as a universal foundation of primary education.

In this chapter, therefore, a quantitative model will be built with the dependent variable *literacy*, which could be an indicator of "human capability." How can this theoretical importance of literacy for human development be translated into practical policy recommendations in a real context? Modeling in this chapter aims at showing (1) an example of a quantitative approach to education policy analysis and (2) a possibility to bridge theoretical consideration and practical policy making. The case of Tanzania, one of the sub-Saharan nations, is focused and analyzed.

Quantitative Consideration on Literacy and Social Service: The Case of Tanzania

A Concept of the Model

As discussed above, *literacy* is taken as the dependent variable. Considering literacy as one of the basic abilities that one should gain at the early stage of life, the model considers school-age children as its sample. Since a domestic gap between urban areas and rural areas is critical in many developing nations, a rural model and an urban model will be developed separately and compared. By paying attention to school-age children and by taking the urban-rural difference into consideration, the model tries to make "the invisible people," such as rural children who are often ignored in the policymaking process, more visible. Specifically, the model aims at answering these questions: How is Tanzanian children's literacy affected by their life circumstances? How do public services contribute to rural and urban children's literacy? Answers to these questions contribute to understanding the educational development policymaking process for the invisibles in Tanzania.

Data and Variables

Data from the Tanzanian Human Resource Development Survey (THRDS) (The World Bank, 1993, 1997; The World Bank and University of Dar es Salaam, 1993) are used in this chapter. THRDS has one of the most comprehensive survey datasets, compared to other survey datasets that contain single level data only. As a policy study, it is inevitable to consider the influence of public services on children's literacy, therefore, district-level data is necessary, not solely individual-level data. THRDS contains both individual and district-level data; in addition, the object of THRDS is "to assess household welfare,... and to evaluate the effect of various government policies on the living conditions of the population" (Grosh and Glewwe, 1995, 2), which is consistent with the purpose of this chapter.

The Dependent variable *literacy* is defined as a dichotomous variable: 1 (literate) and 0 (illiterate). For independent variables, the model considers two different levels: an individual level and a district level, as shown in figure 14.1.

At the individual level, children's educational environment is considered, and at the district level, the availability of public services in children's general life is considered. The model assumes that these two levels contribute to children's literacy development.

Schooling experience (whether the children have attended school during the past twelve months: yes = 1, no = 0) and book possession (whether the children own a book: yes = 1, no = 0) are selected as independent variables at the individual level, assuming that these two variables reflect the children's educational environment. The Number of teachers, primary schools, dispensaries, health staff, national bank, and percentage of the population with access to clean water in the district are selected as independent variables at the district level, assuming that these six variables reflect children's life environment in terms of public services in fields of education, health, and economics. Other variables may also be considered, but there is an inevitable limitation of data availability related to the case under study. Details of the data used in the model are given below in tables 14.1, 14.2, and 14.3.

Hierarchical Linear Modeling (HLM) was developed to analyze multilevel-structured data, or the data collected from different sampling

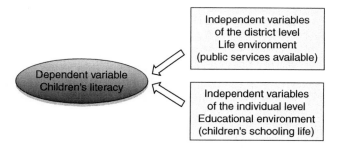

Figure 1 A concept of the model.

Table 14.1   Description of Variables

LEVEL-1 VARIABLES (Individual children aged 7–14)

Literacy	Whether the child can write a letter (yes = 1, no = 0)
Book	Whether the child owns a book (yes = 1, no = 0)
Schooling	Whether the child has attended school during the past 12 months (yes = 1, no = 0)

LEVEL-2 VARIABLES (District)

TCHR	Number of teachers per 10,000 people
PRIM	Number of primary schools per 10,000 people
DISP	Number of dispensaries per 10,000 people
AID	Number of staff for maternal and child health aid per 10,000 people
WAT	Percentage of the population with access to clean water
BRANCH	Number of national bank (community branches) per 10,000 people

Source: Ferreira and Griffin, 1996; World Bank, 1993

Table 14.2   Rural Data

Variable Name	Var Type	N	Frequency of (yes = 1, no = 0)	%
LEVEL-1 DESCRIPTIVE STATISTICS				
Literacy	Dichotomous	(yes = 1, no = 0)	2580	(41,59)
Book	Dichotomous	(yes = 1, no = 0)	2580	(30,70)
Schooling	Dichotomous	(yes = 1, no = 0)	2580	(61,39)

Variable Name	Var Type	N	Mean	SD	Minimum	Maximum
LEVEL-2 DESCRIPTIVE STATISTICS						
TCHR	Continuous	57	49.54	21.62	9.82	125.39
PRIM	Continuous	57	5.22	1.15	2.51	8.23
DISP	Continuous	57	1.25	0.56	0.00	2.49
AID	Continuous	57	1.40	0.56	0.19	3.01
WAT	Continuous	57	36.32	20.98	0.08	100.00
BRANCH	Continuous	57	0.07	0.06	0.00	0.36

Note: Analyzed by HLM 6.02

levels, such as an individual level and a group level that contains the individuals in it (Luke, 2004; Raudenbush and Bryk, 2002; Raudenbush et al., 2004; Snijders and Bosker, 2003). For the analysis in this chapter, the data contain both individual-level and district-level variables. Individual samples in this data are from a specific district; in other words, they are *nested* into a specific district and share the district-level information among other individuals living in the same district. When

Table 14.3   Urban Data

Variable Name	Var Type		N	Frequency of (yes = 1, no = 0)	%
LEVEL-1 DESCRIPTIVE STATISTICS					
Literacy	Dichotomous		(yes = 1, no = 0)	2941	(54,46)
Book	Dichotomous		(yes = 1, no = 0)	2941	(39,61)
Schooling	Dichotomous		(yes = 1, no = 0)	2941	(67,33)

Variable Name	Var Type	N	Mean	SD	Minimum	Maximum
LEVEL-2 DESCRIPTIVE STATISTICS						
TCHR	Continuous	39	58.85	30.44	8.42	160.20
PRIM	Continuous	39	4.37	1.75	0.24	8.96
DISP	Continuous	39	1.80	1.26	0.21	6.01
AID	Continuous	39	2.36	1.93	0.19	9.59
WAT	Continuous	39	41.73	27.84	1.50	100.00
BRANCH	Continuous	39	0.20	0.18	0.02	0.85

Note: Analyzed by HLM 6.02

analyzing this kind of nested or multilevel-structured data, HLM is an appropriate method. In addition, the model in this chapter contains a dichotomous dependent variable *literacy*, which means that the model assumes nonlinearity. HLM with a dichotomous outcome variable is called Hierarchical Generalized Linear Modeling (HGLM) (Raudenbush and Bryk, 2002; Snijders and Bosker, 2003). HGLM has similar characters to logistic modeling in terms of its function, estimation process, and outcome interpretation (Hosmer and Lemeshow, 2000; Long, 1997; Pampel, 2000; Menard, 2002). In this chapter, HGLM is applied for the two-level structured data (individual and district levels) using a dichotomous outcome variable *literacy* to examine how individual-level variables (schooling experience and book possession) and district-level variables (educational, health, and economic public services) contribute to children's literacy. The data include rural samples (n = 2985 including missing, n = 2580 after list-wise deletion; 7-14-year-old school-age children) and urban samples (n = 3062 including missing, n = 2941 after list-wise deletion; 7-14-year-old school-age children) created from THRDS. Technically, the model function is defined as follows.
The individual-level link function:

$$g(x) = \ln = \left[\frac{\pi(x)}{1 - \pi(x)} \right] \beta_{0j} + \beta_{1j} * X_{book} + \beta_{2j} * X_{schooling}.$$

The district-level model:

$$\beta_{0j} = G_{00} + G_{01} * W_{DISP} + G_{02} * W_{AID} + G_{03} * W_{BRANCH} + G_{04}$$
$$* W_{TCHR} + G_{05} * W_{PRIM} + G_{06} * W_{WAT} + U_{0j},$$

$$\beta_{1j} = G_{10} + G_{11} * W_{TCHR} + U_{1j},$$

$$\beta_{2j} = G_{20} + G_{21} * W_{PRIM} + U_{2j}.$$

The combined model:

$$g(x) = G_{00} + G_{01} * W_{DISP} + G_{02} * W_{AID} + G_{03} * W_{BRANCH} + G_{04}$$
$$* W_{TCHR} + G_{05} * W_{PRIM} + G_{06} * W_{WAT} + G_{10} * X_{book} + G_{11}$$
$$* W_{TCHR} * X_{book} + G_{20} * X_{schooling} + G_{21} * W_{PRIM} * X_{schooling}$$
$$+ U_{0j} + U_{1j} * X_{book} + U_{2j} * X_{schooling}$$

$$= \underbrace{\{G_{00} + G_{01} * W_{DISP} + G_{02} * W_{AID} + G_{03} * W_{BRANCH} + G_{04}}_{}$$
$$\underbrace{* W_{TCHR} + G_{05} * W_{PRIM} + G_{06} * W_{WAT}\}}_{Intercept = \beta_{0j}}$$

$$+ \underbrace{\{G_{10} + G_{11} * W_{TCHR}\} * X_{book}}_{book\ slope = \beta_{1j}}$$

$$+ \underbrace{\{G_{20} + G_{21} * W_{PRIM}\} * X_{schooling}}_{schooling\ slope = \beta_{2j}}$$

$$+ \underbrace{\{U_{0j} + U_{1j} * X_{book} + U_{2j} * X_{schooling}\}}_{error}.$$

In this model, β_{0j} (intercept) represents the likelihood of *literacy* without controlling for individuals' conditions of *book possession* and *schooling experience*. G_{00} specifically represents the ground mean likelihood of *literacy*; that is, the mean across all districts and individuals. β_{0j} (intercept) in this model assumes that this general likelihood of *literacy* may be influenced by each district's unique conditions of public services (Ws). Ws in β_{0j} (intercept) are considered as direct conditions contributing to an outcome variable *literacy*. In other words, all public service factors are considered as direct influences on *literacy* because they represent general life circumstances for *all* children without reference to the individual conditions of *schooling experience* or *book possession*.

β_{1j} (*book* slope) represents an *additional* effect on the likelihood of *literacy* of those who own at least one book; in other words, the coefficient $\{G_{10} + G_{11} * W_{TCHR}\}$ will disappear if the individual does not own any books ($\because X_{book} = 0$). β_{1j} (*book* slope) in this model contains a predictor TCHR, assuming that teachers' encouragement to students to read the books may contribute to developing literacy, and that more teachers may provide more chances for children to receive such encouragement. However, the contribution from PRIM to β_{1j} (*book* slope) is not assumed because textbooks are usually not required in Tanzanian primary schools, and it is the responsibility of individual families' to decide to purchase textbooks.[2]

β_{2j} (*schooling* slope) represents another *additional* effect on the likelihood of *literacy* for those who have schooling experience. In other words, the coefficient $\{G_{20} + G_{21} * W_{PRIM}\}$ will disappear if the individuals do not have schooling experience ($\because X_{school} = 0$). β_{2j} (*schooling* slope) in this model contains a district-level predictor PRIM, assuming that more schools in a district may bring children higher accessibility to education, which may contribute to literacy development.[3]

This model indicates two concerns: (1) how do the individual-level conditions (*book* and *schooling*) and the district-level services (six Ws) contribute to the likelihood of children's literacy; and (2) whether there are significant interactions between children's book possession and the number of teachers ($W_{TCHR} * X_{book}$) and between children's schooling experience and the number of primary schools ($W_{PRIM} * X_{schooling}$). This model produces the results presented in table 14.4.

Table 14.4 Final Model

	Coefficient	SE	T-ratio	df	OR	Confidence Interval
FIXED EFFECT						
Rural For Intercept B0j						
Intercept G00	−3.167	0.147	−21.523	50	0.042**	(0.031,0.057)
DISP G01	−0.167	0.141	−1.190	50	0.846	(0.638,1.122)
AID G02	0.459	0.120	3.828	50	1.583**	(1.244,2.013)
BRANCH G03	−1.203	1.778	−0.677	50	0.300	(0.008,10.617)
TCHR G04	−0.005	0.004	−1.223	50	0.995	(0.987,1.003)
PRIM G05	0.220	0.165	1.331	50	1.246	(0.895,1.735)
WAT G06	0.004	0.003	1.245	50	1.004	(0.998,1.010)
Urban For Intercept B0j						
Intercept G00	−3.009	0.160	−18.758	32	0.049**	(0.036,0.068)
DISP G01	0.027	0.067	0.405	32	1.028	(0.896,1.179)
AID G02	0.026	0.041	0.634	32	1.027	(0.944,1.117)
BRANCH G03	0.687	0.541	1.269	32	1.987	(0.661,5.974)

Continued

Table 14.4   Continued

	Coefficient	SE	T-ratio	df	OR	Confidence Interval
TCHR G04	0.002	0.003	0.841	32	1.002	(0.997,1.008)
PRIM G05	−0.124	0.068	−1.829	32	0.883	(0.769,1.014)
WAT G06	−0.006	0.003	−1.905	32	0.994	(0.987,1.000)
Rural For Book slope B1j						
Intercept G10	1.498	0.157	9.533	55	4.472**	(3.266,6.125)
TCHR G11	0.015	0.006	2.401	55	1.015*	(1.002,1.028)
Urban For Book slope B1j						
Intercept G10	1.407	0.144	9.771	2930	4.083**	(3.079,5.414)
TCHR G11	−0.003	0.003	−1.308	2930	0.997	(0.992,1.002)
Rural For Schooling Slope B2j						
Intercept G20	3.189	0.148	21.492	2569	24.261**	(18.139,32.449)
PRIM G21	−0.146	0.174	−0.837	2569	0.864	(0.614,1.216)
Urban For Schooling Slope B2j						
Intercept G20	3.446	0.172	20.011	37	31.380**	(22.149,44.459)
PRIM G21	0.068	0.079	0.856	37	1.070	(0.912,1.255)

	SD	Variance Component	df	Chi-squre
RANDOM EFFECT				
Rural Intercept U0j	0.285	0.081	48	66.959*
Urban Intercept U0j	0.130	0.017	32	26.945
Rural Book slp U2j	0.797	0.635	53	88.852**
Urban Book slp U2j		Constrained		
Rural Schooling slp U2j		Constrained		
Urban Schooling slp U2j	0.336	0.113	37	30.359
Rural Tau as correlations	U0j	U1j		

$$\begin{bmatrix} 1000 & -0.437 \\ & 1.000 \end{bmatrix}$$

Urban Tau as correlations	U0j	U2j

$$\begin{bmatrix} 1000 & -0.633 \\ & 1.000 \end{bmatrix}$$

STATISTICS FOR CURRENT MODEL

	df	Deviance
Rural For the deviance test	14	6951.144
Urban For the deviance test	14	7814.628

Notes:
_1: ** = p < .01; * = p < .05
_2: Method of estimation = EM Laplace
_3: Distribution at level-one = Bernoulli

Note: Analyzed by HLM 6.02 OR = Odds Ratio

The OR in table 14.4 means the "Odds Ratio." Because the outcome variable is dichotomous, coefficients of HGLM are computed as the odds ratios. The OR is a ratio of two odds: the odds of an increase in one additional unit of the relevant predictor over the odds of retaining the existing condition of the relevant predictor. The OR is defined as an indicator that measures "how much more likely (or unlikely) it is for the outcome to be present among those with x = 1 [the event happened: *to be literate* in this model] than among those with x = 0 [the event did not happen: *to be illiterate* in this model]" (Hosmer and Lemeshow, 2000, 50). The OR can take the minimum value of zero and go up to a maximum of infinity. In other words, the OR indicates how many times the outcome likelihood changes in relation to changes in the relevant predictor by one unit (Hosmer and Lemeshow, 2000, 48–56; Long, 1997, 79–82; the American College of Physicians, 2000, 145–146). Therefore, the OR = 1 means no change; a value of the OR > 1 indicates an increased likelihood; and a value of the OR < 1 implies a decreased likelihood.

The idea of the OR for literacy is not yet widely accepted as a policy criterion compared to the use of the literacy ratio. Nonetheless, the OR can be called a more individual-focused indicator than the literacy ratio. The Literacy ratio represents the proportion of literate people in the total population while The OR for literacy is concerned with changes in the likelihood of literacy for individuals living under certain conditions, which are measured by explanatory variables. The OR provides information on not simply a collective condition of some group-like literacy ratio, but on individual conditions in specific contexts.

Interpretation of the Model

What do the final rural and urban models indicate about children's literacy in Tanzania? In a comparison of the rural and urban models (see table 14.4), the individual-level variables (*schooling experience* and *book possession*) show some similarities: children who own at least one book are about four times as likely to be literate as those who do not own any books, and children who have schooling experience are about twenty-four times (in rural areas) and thirty-one times (in urban areas) as likely to be literate as those who have no schooling experience. These results verify the positive effects of book possession and primary schooling for children's literacy development, supporting Elley's (2000) "Book Flood approach" and other literature that claims the significance of schooling for children's literacy development (Haidara, 1990; Muller and Murtagh, 2002). Moreover, these findings endorse Samoff's (1990) statement that schooling plays a central role in Tanzanian education. On the other hand, the district-level variables yield different results for the rural and urban models. Specifically, the district-level variables make significant contributions to children's literacy in the rural districts but not in the urban districts.

How does the rural model speak to human development policy? In the rural model, the number of health aid staff (AID) and the interaction of book possession and the number of teachers (*book**TCHR) showed significant contributions to children's literacy (see table 14.4). Reflecting existing conditions in rural Tanzania, the findings from the model can be utilized for policy recommendations. Specifically, it verifies the important effects of health aid staff and teacher-book interactions on the likelihood for a rural Tanzanian child to be literate. The details are discussed below.

a.   Needs for health aid staff

The existing conditions of public health services in Tanzania are shown in table 14.5.

The number of health staff per health facility shows that less than one staff member is employed for a typical health facility in a typical rural district (0.914 per health facility), while a typical urban health facility employs on average at least one health aid staff member. Although the data above are quite rough since the values are means of all rural or urban districts, it is clear that the condition of rural health

Table 14.5   District Means: Health Services

	Rural	*Urban*
Mean of population:	273713	315320
Mean of number if dispensaries:	30	34
Mean of number of hospitals:	2	6
Mean of number of health staff:	32	50
Mean of number of doctors:	2	17
Mean of the district area (km^2):	10128	2483
(From THRDS, World Bank, 1993)		
(1) Number of dispensaries per 10km^2:	0.030	0.137
(2) Number of hospitals per 10km^2:	0.002	0.024
(3) Number of health staff per health facility*:	0.914	1.136
(4) Number of doctors per hospital:	1.000	2.833

Note: *Either a dispensary, hospital, or health center.
Source: (Calculated by the author)

services is poorer than it is in the urban areas. Reflecting the insufficient condition of health services in rural districts and the statistical significance of AID in the model, it can be reasonable to consider AID or health aid staff as a part of literacy development policy for rural children.

Policy-makers may need to know the anticipated effect of a policy in advance in order to decide if that particular policy is worth implementing. This quantitative model works as a needs assessment model to make the voice of those who are often invisible louder. The model cannot assume causation; however, the OR of AID indicate a positively significant contribution to rural children's literacy.

b.   Needs for books and teachers

Table 14.6 summarizes the existing conditions of public educational services in Tanzania.

The mean number of primary schools in the rural districts (133) shows a higher value than that for the urban districts (82). However, adjusting for the differences in district sizes, the rural mean of the number of primary schools per 100 square kilometers works out to 1.313, while the urban mean is 3.302. Therefore, in terms of accessibility to school, rural children face greater challenges than urban children. Similarly, the rural mean of the number of teachers (1191) is slightly higher than the urban mean (1153). However, considering the difference in the number of primary schools between rural and urban areas, the rural mean of the number of teachers per school is 8.955, while the urban mean is 14.061. Therefore, in terms of accessibility to teachers, the rural children face more difficult circumstances than the urban children. Although the data above are quite rough since they are mean

Table 14.6   District Means: Educational Services

	Rural	*Urban*
Mean of population:	273713	315320
Mean of number of primary schools:	133	82
Mean of number of teachers:	1191	1153
Mean of the district area (km²):	10128	2483
(1) Number of primary schools per 100km²:	1.313	3.302
(2) Number of teachers per school:	8.955	14.061

Source: From THRDS, World Bank, 1993; and calculated by the author

values of all rural or urban districts, it is clear that the condition of educational services in rural areas is poorer than the condition in urban areas.

The claim that more teachers are needed in the rural districts seems to be reasonable, but it should be noted that *book* * TCHR is an *interaction* term. Since this is an interaction term, this positive effect can be expected only for children who own at least one book (*book* = 1). For those children who do not possess any books (*book* = 0), this interaction term disappears from the model. Although the model cannot assume causation, there may be two different ways of interpreting this interaction term.

1. Teachers encourage children to own and read books. Therefore, having more teachers indicates increasing opportunity for children to be encouraged to read books. This may be why the interaction term positively contributes to children's literacy.
2. Children who own books are those from wealthy or educated families. Therefore, those children are amenable to education and will certainly acquire literacy if there are teachers available to them. This may be why the interaction term positively contributes to children's literacy.

In the case of (1), the recommendation of creating more teachers can be justified; in the case of (2), the significant factor is books rather than teachers, and a recommendation for free textbook distribution may be a more reasonable one. The model does not imply causation, but it can be said that both books and teachers are important factors for improving rural children's literacy. Here again, the model contributes to making the needs of those who are often invisible more visible to policy-makers.

Making the Invisible Visible

This chapter showed an example of quantitative modeling for educational policy with an analysis of Tanzania's data. The method, HGLM, clarified what factors affect Tanzanian children, especially rural children, in terms of their literacy development. Although there may be a countless number of other issues related to literacy development in Tanzania, the model provides a possible direction for it. HGLM enabled analysis of their circumstances from two different levels: the district level and the individual level. This approach would help policy-makers reconsider the need for public services for rural children who are often ignored in the policy context. Although the findings here may not be robust enough to be directly applied to a real policy, they have shown some examples of policy implications from a standpoint of rural school-age children in Tanzania. Further field research, including qualitative research, is necessary to determine the details of the situation. However, quantitative research could provide guidance to identify and clarify invisible problems and/or invisible people in the policy context and could describe their existence within a specific form by using models, indices, and their values.

A quantitative approach in social science does not always mean that its findings are objective or neutral. When choosing variables, when modeling, and when interpreting results, subjectivity and bias are invariably present. However, this approach can help us make invisible aspects, such as the needs of voiceless people, more visible. In the case of developing nations where contexts and structures are complex and often unknown, this becomes particularly valuable. Ackoff (1974) states that "[s]uccessful problem solving requires finding the right solution to the right problem. We fail more often because we solve the wrong problem than because we get the wrong solution to the right problem." Educational research needs to contribute to finding "the right problem" that is worth solving, and "the right problem" could be invisible without careful consideration. If this is the case, it is the educational researchers' first job to make it visible.

Notes

1. This chapter is a reedited version of the dissertation: Yonehara, A. (2006). *Human development policy: Theorizing and modeling.* Indiana, Bloomington.
2. PRIM and BRANCH were also tested by Wald tests, and they showed insignificance.
3. BRANCH and TCHR were also tested by Wald tests, and they showed insignificance.

Methods and the Margins: Realigning the Center, A Postscript

Debora Hinderliter Ortloff,
Cheryl A. Hunter, and Rachelle Winkle-Wagner

The idea for this book started as a conversation between friends in Barbara Dennis' living room. The editors, together with Barbara Dennis, Adrea Lawrence, and Joshua Hunter, had come together to discuss our work and broadly some of the trends in educational research. It became an extended, if not unending, conversation. We were mindful of our elite position; in fact, one of the key points in our discussion was our relative security in the center. We were White, Western women and men who had not only the luxury of education, but the indulgence of time—time to think and explore new ideas and innovations. Yet, each of us shared stories of marginalization, because of our gender, our sexuality, our age, our reproductive decisions, our geography, and universally our intellectual work. In this latter category, all of our experiences seemed to mesh. It was the experience of marginalization that had led us to investigate certain topics in our research: birth education, educational experiences of marginalized groups in and outside of the United States, and the role of natural objects in educative experiences, to name a few.

On the surface, even as we write these topics, they seem to raise important, interesting, and relatively "normal" or "mainstream" questions about education and educative experiences. And yet producing this work removed us farther from the center when that work was deemed edgy or unconventional, or as one reviewer noted, "too feminist for our [feminist] journal." Trained in critical theory, and specifically critical race theory, we were already very keyed into ideas of power and marginality, particularly as it relates to identity and privilege. However, we began, in those same conversations around the coffee table, to explore the notion of the margins as it relates to the production of educational research. We questioned why, in any given educational journal, similar topics seem to be hashed and rehashed from one year to the next. Clearly certain groups were underrepresented, if not absent, as both subjects and producers of research. But beyond this, it seemed also that, although good research was clearly being done, the research, theoretically and empirically, seemed to show very little variation in the methodologies, reporting of findings, and often in the topics or people being studied. We began to think about the connection between the absence of research by and about marginalized groups and the methodological tools available. We asked ourselves: What ways of knowing are needed to discover the margins or to bring the margins to the center? In discussing this, we realized these questions required a more fundamental question to be first considered: What are the margins?

And so our living room chats, which had now gone on for four years, through graduations, promotions, births, deaths, and relocations, continued. We also invited more people to our conversations, as evidenced by the chapters in this book. What we discovered, as we have shared throughout this book, was a much broader concept of marginality—for while there may be a clear delineation between center and margin, mainstream and "edgy," there are no edges to the margins. As educational researchers, we have to constantly challenge ourselves to rethink what constitutes education, what "counts" as research, what ways and means we have and need to have, to discover new theoretical, empirical, and methodological ideas about education. This challenge requires thoughtful, intentional reflection at every level of the research process. It also requires the willingness to call out and challenge the hierarchy of the status quo and our position in its reproduction, particularly as members of the academic research community. As noted in the introduction of this volume, "we are not the first to highlight marginality." The persistence of the center and the margins was our primary impetus for beginning this project, and its amelioration remains our chief concern. Our contribution to this endeavor is a commitment to transformative research.

We have tried throughout this book to show that marginality is a concept that can be, when left unchecked and unprobed, defining for research. What we mean here is that the center/margin divide influences all aspects of the research process, who researches, what gets asked to whom and by whom, what gets funded, what gets published, what research methods are used, and what data are valued. We argue that marginality in research is more than the absence of particular voices from research. Indeed, the work here shows that research aimed at working from a different perspective, one that opposes the center/margin divide, must conceive of marginality as paradigmatic and pervasive in the research process in order to overcome it. Marginality in research exists because the system structures that create space for research (e.g., political structures, institutional structures, social structures) interact to keep people, ideas, groups, innovations, and so on from becoming part of the mainstream research process. As Apple points out in the beginning of this book, the relationships between political, economic, and cultural spheres cannot be divorced from education or educational research. In other words, the research process itself decenters groups and ideas continually. This more expansive idea of marginality necessarily means that giving voice to underrepresented groups, as has been argued by a variety of social justice oriented researchers (see, for example, the works of Ladsen-Billings, 2001; Friere, 1986; Fine, 2007; Giroux, 2008; hooks, 2004), is only one important step in removing the center/margin divide. In this book, then, we aimed instead at addressing the idea of marginality in research in terms of theory, institutional practice, methodology, and by providing examples of work that addresses the margins. By doing this, we are answering the assertion in the introduction that by allowing the margins to inform the center we are able to expand knowledge by explicating ways of knowing (theoretically, methodologically, and empirically) that have been historically unexplored. Theory, methodology, data collection, and analysis must all consider and reconsider how they have the potential to marginalize (or not marginalize), but this must take place within a democratic and fair research system in order for it to be effective. Using methods on the margins and looking at "marginal" cases have power because they interrupt what is expected, potentially transforming research practices and knowledge more generally.

This is the interest of this book—to contemplate change and newness in educational research. By doing this, we call into question any research process that reifies the center/margin divide. We hope, and it is our intention, also to call out to others to journey out of the center. Delandshere's piece, in the first section of this volume, discusses the political climate surrounding educational research in the United States, and throughout this book several contributors as well as ourselves have discussed how institutions and processes contribute to the continuation of this dialectic. Academic processes promote safe, rather than innovative, research, particularly in the social sciences where "new" and "discovery" are more relative terms. Educational research, in particular, remains tied to concepts, as Lawrence points out, of education as schooling. Tenure and promotions procedures favor non-collaborative work that is antithetical to the embracing of democratic processes for research. The process of creating a hierarchical scale upon which to be measured in the academy reifies the hegemony; only what has already been recognized as valid and important can be valued, so there is no space for decentered knowledge. This academic research tradition, as Smith (2008) points out, furthers the divide between theory and practice by promoting research that does not recognize educators (as opposed to educational researchers) as equals. These particulars create the structures in which educational research exists and by which the center/margin divide is reproduced as each subsequent generation of educational researchers comes to the helm.

What then do the arguments in this volume mean for us as researchers? We hope it, at the very least, provides a beginning. For those scholars who want to research their passions but do not see a place for this passion in the annals of research, we hope the work presented here shows a way and imparts energy to those who may feel they just keep hitting their head against the wall. We fear that adherence to traditional scholarship without question means we are losing fantastic scholars, and important, often disenfranchised voices, because they just cannot find the energy anymore. Our aspiration in presenting the margins of methodology is that others will take up this work and continue it.

Beyond this, it is our intention to call out to others because overcoming the margins requires more than the individual action of the researcher, as they consider the theory, methodology, and empirical implications of their own work. It requires a concerted effort of educational researchers to reshape, as a field, our understanding and acceptance of the center/margin divide. In the introduction, we called on the words of the Nobel Prize Laureate, Elie Wiesel (1986). Wiesel's words that we read initially, at those first coffee klatches six years ago, speak of making the fight for injustice at the very center. He further notes:

> As long as one dissident is in prison, our freedom will not be true. As long as one child is hungry, our life will be filled with anguish and shame. What all these victims need above all is to know that they are not alone; that we are not forgetting them, that when their voices are stifled we shall lend them ours, that while their freedom depends on ours, the quality of our freedom depends on theirs.

Here Wiesel is reminding us that in all that we do we must consider ourselves as part of the margins so as to never let those marginalized go unheard. We interpret this to mean that, as researchers, we must dedicate and rededicate ourselves to research that seeks to transform. This research, we believe, must not just give voice to those marginalized, or speak for those who cannot, as Wiesel charges, but create a space for the concerns, experiences, and values of the margins to be incorporated and interpreted on their own terms. This requires us to step outside of the research process as it is taught and retaught in the halls of academia and explore new ways of knowing. At each step of the research process, both as individuals and as members of institutions that provide structures and rewards for research, we have to check the margin/center divide and work to remove it. We can do no less.

References

Abbott, E. H. (1903). The fair at St. Louis. *Outlook (1893–1924)*, 74(10), 552–563.

Ackoff, R. L. (1974). *Redesigning the future: A systems approach to societal problems*. New York: Wiley.

Adams, D. W. (1995). *Education for extinction: American Indians and the boarding school experience*. Lawrence: University Press of Kansas.

———. (2006). Land, law, and education: The troubled history of Indian citizenship,1871–1924. In D. Warren & J. J. Patrick (Eds.), *Civic and moral learning in America* (119–134). New York: Palgrave & Macmillan.

Agency for Healthcare Research and Quality. (2006). *Program brief: Research on cardiovascular disease in women*. Rockville, MA: U.S. Department of Health and Human Services. Accessed March 1, 2007 from http://www.ahrq.gov/research/womheart.pdf.

Anderson, B. (1991). *Imagined Communities: Reflections on the origin and spread of nationalism*. London: Verso.

Apple, M. W. (1996). *Cultural politics and education*. New York: Teachers College Press.

———. (1999). *Power, meaning, and identity*. New York: Peter Lang.

———. (2000). *Official knowledge* (2nd ed.). New York: Routledge.

———. (2004). *Ideology and curriculum* (3rd ed.). New York: Routledge.

———. (2005). *Education and power*. New York: Routledge.

———. (2006a). *Educating the "right" way: Markets, standards, God, and inequality* (2nd ed.). New York: Routledge.

———. (2006b). Rhetoric and reality in critical educational studies in the United States. *British Journal of Sociology of Education*, 27, 679–687.

Apple, M. W. & Weis, L. (Eds.). (1983). *Ideology and practice in schooling*. Philadelphia: Temple University Press.

Apple, M. & Aasen, P. (2003). *The state and politics of education*. New York: Routledge.

Apple, M. W. & Beane, J. A. (Eds.). (2007). *Democratic schools: Lessons in powerful education*. Portsmouth, NH: Heinemann.

Apple, M. W. & Buras, K. L. (Eds.). (2006). *The subaltern speak: Curriculum, power, and educational struggles*. New York: Routledge.

Arbona, C. & Nora, A. (2007). Predicting college attainment of Hispanic students: Individual, institutional, and environmental factors. *The Review of Higher Education*, 30(3), 247–270.

Attinasi, L. C., Jr. (1989). Getting in: Mexican Americans' perceptions of University attendance and the implications for freshman year persistence. *Journal of Higher Education*, 60, 247–277.

Attinasi, L. C., Jr. & Nora, A. (1987). The next step in the study of student persistence in college. Paper presented at the meeting of the Association for the Study of Higher Education, Baltimore, MD.

Au, W. & Apple, M. W. (2007). Freire, critical education, and the environmental crisis. *Educational policy*, 21, 457–470.

Augustine, S. (1991). *Saint Augustine confessions*. Oxford: Oxford University Press.

Avellaneda, I. (1997). Hernando de Soto and his Florida fantasy. In P. Galloway (Ed.), *The Hernando de Soto expedition* (207–218). Lincoln: University of Nebraska Press.

Axtell, J. (2001). *Natives and newcomers: The cultural origins of North America*. New York: Oxford University Press.

Bailyn, B. (1960). *Education in the forming of American society: Needs and opportunities for study*. New York: Vintage Books.

Bakhtin, M. M. (1968). *Rabelais and his world* (Trans. H. Iswolsky). Cambridge, MA: M.I.T. Press.

Banks, J. A. (1998). The lives and values of researchers: Implications for education citizens in a multicultural society. *Educational Researcher*, 27(7), 4–17.

———.(Ed.). (2004). *Diversity and citizenship education*. San Francisco: Jossey-Bass.

———. (April, 2008). Diversity, group identity, and citizenship education in a global age. *Educational Researcher*, 37(3), 129–139.

Basso, K. (1996). *Wisdom sits in places: Landscape and language among the Western Apache*. Albuquerque, NM: University of New Mexico Press.

Behar, R. (1993). *Translated woman: Crossing the border with Esperanza's story*. Boston: Beacon.

Benhabib, S. (1992). *Situating the self: Gender, community and postmodernism in contemporary ethics*. Routledge: New York.

Bensimon, E., Nora, A., Patriquin, L., & LaBare, M. (1999, November). *A comparison of qualitative and quantitative findings among college students*. Symposium conducted at the annual meeting of the Association for the Study of Higher Education.

Benton, T. & Craib, I. (2001). *Philosophy of social science: The philosophical foundations of social thought*. New York, NY: Palgrave.

Bhabha, H. (1994). *The location of culture*. New York: Routledge.

Blackmon, D. A. (2008). *Slavery by another name: The re-enslavement of Black Americans from the Civil War to World War II*. New York: Doubleday.

Bolton, C. D. & Kammayer, K. C. W. (1972). Campus cultures, role orientations, and social types. In K. Fledman (Ed.), *College and student: Selected Readings in the social psychology of higher education.* New York: Pergamon.

Boruch, R. F. (1982). Experimental tests in education: Recommendations from the Holtzman report. *The American Statistician,* 36(1), 1–8.

Boruch, R. F. and Cordray, D. S. (Eds.). (1982). *An appraisal of educational program evaluation: Federal, state, and local agencies.* New York: Cambridge University Press.

Bourdieu, P. (1984). *Distinction: A social critique of the judgment of taste.* (Trans. Richard Nice) Cambridge, MA: Harvard University Press.

———. (1988). *Homo academicus.* Stanford: Stanford University Press.

———. (2003). *Firing back: Against the tyranny of the market 2.* New York: New Press.

Bowers, C. A. (1993a). *Critical essays on education, modernity, and the recovery of the ecological imperative.* New York: Teachers College Press.

———. (1993b). *Education, cultural myths, and the ecological crisis: Toward deep changes.* Albany, NY: SUNY Press.

Brandom, R. B. (1994). *Making it explicit; Reasoning, representing, and discursive commitment.* Cambridge, MA: Harvard University Press.

Bridges, D. The ethics of outsider research. *Journal of Philosophy of Education,* 35(3), 371–386.

Bruess, C. E. & Greenberg, J.S. (2004). *Sexuality education: Theory and practice.* New York: Macmillan Publishing.

Bruner, E. (1986). Experience and Its Expressions. In *The anthropology of experience* (3–30). Urbana and Chicago: University of Illinois Press.

Burawoy, M. (2005). For public sociology. *British Journal of Sociology of Education,* 56, 259–294.

Burbules, N. C. & Torres, C. A. (Eds.). (2000). *Globalization and education: Critical perspectives.* New York, NY: Routledge.

Butler, J. (1990). *Gender trouble: Feminism and the subversion of identity.* New York: Routledge.

Cabrera, A. F., Nora, A., Terenzini, P. T., Pascarella, E. T., and Hagedorn, L. S. (1999). Campus racial climates and the adjustment of students to college: A comparison between white students and African American students. *Journal of Higher Education,* 70(2), 134–160.

Carnevale, A. P. & Desrochers, D. M. (2003). *Standards for what? The economic roots of K-16 reform.* Princeton, NJ: Educational Testing Service.

Carspecken, P. F. (1996). *Critical ethnography in educational research: A theoretical and practical guide.* New York: Routledge.

———. (1999). *Four scenes for posing the question of meaning and other essays in critical philosophy and critical methodology.* New York: Peter Lang.

———. (2001). Critical ethnographies from Houston: Distinctive features and directions. In P. F. Carspecken and G. Walford (Eds.), *Critical ethnography and education: Vol. 5. Studies in educational ethnography.* Amsterdam, New York, Oxford: JAI Press.

Carspecken, P. F. (2003). Ocularcentrism, phonocentrism and the counter-enlightenment problematic: Clarifying contested terrain in our schools of education. *Teachers College Record,* 105(6), 978–1047.

———. (2006). Limits of knowledge in the physical sciences. In K. Tobin and J. Kincheloe (Eds.), *Doing educational research: A Handbook* (405–438). Rotterdam: Sense Publishers.

Casey, E. S. (1993). *Getting back into place: Toward a renewed understanding of the place-world.* Bloomington, IN: Indiana University Press.

———. (1996). How to get from space to place in a fairly short stretch of time: Phenomenological prolegomena. In S. Feld & K. H. Basso (Eds.), *Senses of place.* Santa Fe, NM: School of American Research Press.

———. (1997). *The fate of place: A philosophical history.* Berkeley, CA: The University of California Press.

Castellanos, J. & Jones, L. (2004). Latino/a undergraduate experiences in American higher education. In J. Castellanos & L. Jones (Eds.), *The majority in the minority.* Sterling, VA: Stylus.

Charmaz, K. (2006). *Constructing grounded theory: A practical guide through qualitative analysis.* London: Sage Publications.

Clifford, J. & Marcus, G. E. (Eds.). (1986). *Writing culture: The poetics and politics of ethnography.* Berkeley: University of California Press.

Conn, S. (2004). *History's shadows: Native Americans and historical consciousness in the nineteenth century.* Chicago: University of Chicago Press.

Corle, E. (1972 [1935]). *Fig tree John.* New York: Pocket Books.

Correspondence. (1903). *Outlook (1893–1924),* 75(9), 519.

Corsaro, W. A. (2005). *Sociology of childhood.* Thousand Oaks: Pine Forge Press.

Counts, G. (1932). *Dare the school build a new social order?* New York: John Day Company.

Crandall, C. (1904a). Letter to Clara D. True regarding the potter going to the St. Louis Expo, January 22, 1904. *Press copies of letters sent concerning pueblo Day Schools.*

———. (1904b). Letter to Clara D. True regarding potters going to the St. Louis Expo, February 15, 1904. *Press copies of letters sent concerning Pueblo Day Schools.*

———. (1904c). Letter to Cora Marie Arnold regarding Albino Chavarria's land problems, February 29, 1904. *Press copies of letters sent ("Miscellaneous Letters").*

———. (1904d). Letter to Pedro Cajete regarding recruiting Pueblo students to the Chilocco Indian School, June 25, 1904. *Press copies of letters sent ("Miscellaneous Letters").*

Cremin, L. A. (1965). *The wonderful world of Ellwood Patterson Cubberley: An essay on the historiography of American education.* New York: Bureau of Publications, Teachers College, Columbia University.

Cronbach, L. J., Ambron, S. R., Dornbusch, S. M., Hess, R. D., Hornik, R. C., Phillips, D. C., et al. (1980). *Toward reform of program evaluation*. San Francisco: Jossey-Bass.

Danziger, K. (2000). Making social psychology experimental: A conceptual history, 1920–1970. *Journal of the History of the Behavioral Sciences*, 36(4), 329–347.

Darnton, R. (1982). *The literary underground of the old regime*. Cambridge, MA: Harvard University Press.

Davis, M. (2006). *Planet of slums*. New York: Verso.

Deagan, K. A. (1985). Spanish-Indian Interaction in sixteenth-century Florida and Hispaniola. In W. W. Fitzhugh (Ed.), *Cultures in contact* (281–318). Washington, DC: Smithsonian Institution Press.

Delandshere, G. (2004). The moral social and political responsibility of educational researchers: Resisting the current quest for certainty. *International Journal of Educational Research*, 41(3), 237–256.

Delandshere, G. & Petrosky, A. R. (1994). Capturing teachers' knowledge: Performance assessment a) and post-structuralist epistemology, b) from a post-structuralist perspective, c) and post-structuralism, d) none of the above. *Educational researcher*, 23(5), 11–18.

Deloria, V., Jr. (1991a). Foreword. In D. A. Grinde, Jr. & B. E. Johansen (Eds.), *Exemplar of liberty: Native America and the evolution of democracy* (ix–xi). Los Angeles: American Indian Studies Center.

———. (1991b). *Indian education in America*. Boulder, CO: American Indian Service and Engineering Society.

———. (1997). *Red earth, white lies: Native Americans and the myth of scientific fact*. Golden, CO: Fulcrum Publishing.

———. (2003). *God is red: A Native view of religion* (30th Anniversary ed.). Golden, CO: Fulgrum Publishing.

———. (2004). Marginal and sub-marginal. In D. A. Mihesuah & A.C. Wilson (Eds.), *Indigenizing the academy: Transforming scholarship and empowering communities* (16–30). Lincoln, NE: University of Nebraska Press.

Denzin, N. K. & Lincoln, Y. S. (Eds.). (2003). Introduction: The discipline and practice of qualitative research. In N. K. Denzin & Y. S. Lincoln (Eds.), *The landscape of qualitative inquiry: Theories and issues* (2nd ed.). Thousand Oaks, CA: Sage Publications, 1–45.

De Tocqueville, A. (2000). *Democracy in America*. (Trans. H. C. Mansfield & D. Winthrop). Chicago: University of Chicago Press. (Original work published in two volumes: volume 1 published in 1835; volume 2 published in 1840).

Dilthey, W. (2002). *Selected works, Volume III; The formation of the historical world in the human sciences*. Princeton and Oxford: Princeton University Press.

Dimitriadis, G. & McCarthy, C. (2001). *Reading and teaching the postcolonial*. New York: Teachers College Press.

Dougherty, K. J. & Hammack, F. M. (1990). Teachers, teaching, and the organizational dynamics of education in the United States. In K. J. Dougherty & F. M. Hammack (Eds.), *Education and society* (168–190). San Diego: Harcourt Brace Jovanovich.

DuBois, W. E. B. (2003). *The souls of Black folk.* New York: The Modern Library. (Original work published 1903).

Dummett, M. (1991). *Frege: Philosophy of mathematics.* Cambridge, MA: Harvard University Press.

Eckardt, N. (2007, November 8). The prevalence of qualitative methodology at AERA's annual meeting and the potential consequences. *Teachers College Record*, Article 14741. Retrieved February 21, 2007, from http://www.tcrecord.org/content.asp?contentid=14741

Educational Progress of the Year. (1904). *Outlook (1893–1924)*, 776.

Eisenhart, M. & Towne, L. (2003). Contestation and change in national policy on "scientifically based?" education research. *Educational Researcher*, 32(7), 31–38.

Eisner, E. (1991). *The enlightened eye: Qualitative inquiry and the enhancement of educational practice.* New York: Macmillan.

Elley, W. (2000). The potential of book floods for raising literacy levels. *International Review of Education*, 46(3/4), 233–255.

Ellis, J. J. (2007). *American creation: Triumphs and tragedies at the founding of the republic.* New York: Alfred A. Knopf.

Ewen, C. R. & Hann, J. H. (1998). *Hernando de Soto among the Apalachee: The archaeology of the first winter encampment.* Gainesville: University Press of Florida.

Feld, S. & Basso, K. (Eds.). (1996). Introduction. In *senses of place.* Santa Fe, NM: School of American Research Press.

Ferreira, M. L. & Griffin, C. C. (1996). *Tanzania human resource development Survey: Final report.* Washington, DC: The World Bank.

Fichte, J. G. (1970). *The science of knowledge.* New York: Cambridge University Press.

Fine, M. (1994). Working the hyphens: Reinventing self and other in qualitative research. In N. K. Denzin & Y. S. Lincoln (Eds.), *Handbook of Qualitative Research* (70–82). Thousand Oaks, CA: Sage Publications.

Fine, M., & Sirin, S. R. (2007). "Hyphenated selves: Muslim American youth negotiating identities on the fault lines of global conflict." *Applied Developmental Science, 11*(3), 151–163.

Fine, M., Weis, L. Weseen, S., & Wong, L. (2000). For whom? Qualitative research, representations and social responsibilities. In N. K. Denzin & Y. S. Lincoln (Eds.), *Handbook of qualitative research* (2nd ed., 107–132). Thousand Oaks, CA: Sage Publications.

———. (2003). For whom? Qualitative research, representations, and social responsibilities. In N.K. Denzin and Y.S. Lincoln (Eds.), *The landscape of qualitative inquiry: Theories and issues* (2nd ed., 167–207). Thousand Oaks, CA: Sage Publications.

Fitzhugh, W. W. (1985). Commentary on Part IV. In W. W. Fitzhugh (Ed.), *Cultures in contact: The impact of European contacts on Native American cultural institutions—A.D. 1000–1800* (271–279). Washington, DC: Smithsonian Institution Press.

Fortman, J. (2003). Adolescent language and communication from an intergroup perspective. *Journal of Language and Social Psychology, 22*(1), 104–111.

Fowler, L. (1996). The Great Plains from the arrival of the horse to 1885. In B. G. Trigger & W. E. Washburn (Eds.), *The Cambridge history of the Native peoples of the Americas* (Vol. 1, Part. 2, 1–55). New York: Cambridge University Press.

Fraenkel, J. & Wallen, N. (2002). *How to design and evaluate research in education with student CD, workbook and powerweb: Research methods* (5th ed.). Columbus, OH: McGraw Hill.

Fraser, N. (1989). *Unruly practices: Power, discourse, and gender in contemporary social theory.* Minneapolis: University of Minnesota Press.

———. (1997). *Justice interruptus.* New York: Routledge.

Freire, P. (1970). *Pedagogy of the oppressed.* New York: Herder and Herder.

Friere, P. (1986). *Pedagogy of the oppressed.* New York: Continuum International Publishing Group.

———. (2000). *Pedagogy of the oppressed* (30th Anniversary ed.). Boston: Continuum International Publishing Group. (Original work published 1977).

Fry, R. (2004). *Latino youth finishing college: The role of selective pathways.* Retrieved June 24, 2004 from the Pew Hispanic Center Web site: www.pewhispanic.org

Fry, G., Chantavanich, S., & Chantavanich, A. (1981). Merging quantitative and qualitative research techniques: Toward a new research paradigm. *Anthropology and Education Quarterly, 12,* 145–148.

Gadamer, H. G. (1989). *Truth and method* (2nd rev. ed., Trans. J. Weinsheimer and D. G. Marshall). New York: Crossroad.

Gaither, M. (2003). *American educational history revisited: A critique of progress.* New York: Teachers College Press.

Galloway, P. (1997a). *Conjoncture* and *longue durée*: History, anthropology, and the Hernando de Soto expedition. In P. Galloway (Ed.), *The Hernando de Soto expedition: History, historiography, and "discovery" in the southeast* (283–294). Lincoln: University of Nebraska Press.

———. (Ed.). (1997b). *The Hernando de Soto expedition: History, historiography, and "discovery" in the southeast.* Lincoln: University of Nebraska Press.

Garrison, J. B. (2006). Ontogeny recapitulates savagery: The evolution of G. Stanley Hall's adolescent. Unpublished doctoral dissertation, Indiana University, Bloomington, IN.

Gasche, R. (1986). *The tain of the mirror: Derrida and the philosophy of reflection.* Cambridge, MA: Harvard University Press.

Geertz, C. (1988). *Works and lives: The anthropologist as author.* Stanford, CA: Stanford University Press.

Geertz, C. (1996). Afterword. In S. Feld & K. H. Basso (Eds.), *Senses of place*. Santa Fe, NM: School of American Research Press.

Giddens, A. (1979). *Central problems in social theory*. Berkley: University of California Press.

————. (1981). *A contemporary critique of historical materialism*. Berkley: University of California Press.

Gilligan, C. (1977). In a different voice: Women's conceptions of self and morality. *Harvard Educational Review*, 47, 481–517.

Giroux, H. A. (2008). The militarization of U.S. higher education after 9/11 (Explorations in critical social science). *Theory, Culture & Society*, 25(5), 56–82.

Gloria, A. M., Castellanos, J., Lopez, A. G., & Rosales, R. (2005). An examination of academic nonpersistence decisions of Latino undergraduates. *Hispanic Journal of Behavioral Sciences*, 27(2), 202–223.

Gonzalez, K. P., Stoner, C., & Jovel, J. (2003). Examining the role of social capital in access to college for Latinas: Toward a college opportunity framework. *Journal of Hispanic Higher Education*, 2, 146–170.

Grady, D., Chaput, L., & Kristof, M. (2003a). Results of systematic review of research of diagnosis and treatment of coronary heart disease in women (Evidence Rep./Tech. Assess.No. 40). Prepared by the University of California San Francisco-Stanford University Evidence-based Practice Center. Contract Number 290-97-0013. Rockville, MD: Agency for Healthcare Research and Quality.

————. (2003b, May). Diagnosis and treatment of coronary heart disease in women: Systematic reviews of evidence on selected topics (Evidence Rep./Tech. Assess. No. 81). Prepared by the University of California, San Francisco-Stanford Evidence-based Practice Center under Contract No 290-97-0013. (AHRQ Publication No. 03-E037). Rockville, MD: Agency for Healthcare Research and Quality.

Gramsci, A. (1971). *Selections from the prison notebooks* (Trans. Q. Hoare and G. N. Smith). New York: International Publishers.

Grant, L. (1993). Gender roles and status in children's peer interactions. *Western Sociological Review*, 18(1), 58–76.

Gratz *v.* Bollinger, 539 U.S. 244 (2003).

Greendorfer. (1991). Gender role stereotypes and early childhood socialization. In G. L. Cohen (Ed.), *Women in sport: Issues and controversies* (3–14). Newbury Park, CA: Sage Publications.

Greenwood, D. J. & Levin, M. (2003). Reconstructing the relationships between universities and society through action research. In N. K. Denzin & Y. S. Lincoln (Eds.), *The landscape of qualitative inquiry: Theories and issues* (2nd ed., 131–166). Thousand Oaks, CA: Sage Publications.

Greymorning, S. (2004). *A will to survive: Indigenous essays on the politics of culture, language and identity*. Boston, MA: McGraw Hill.

Grosh, M. & Glewwe, P. A. (1995). *A guide to living standards surveys and their data sets* (No. 120). Washington, DC: The World Bank.

Grutter *v.* Bollinger, 539 U.S. 306 (2003).

Gunaratnam, Y. (2003). *Researching race and ethnicity: Methods, knowledge and power.* Thousand Oaks, CA: Sage Publications.

Gutmann, A. (1987). *Democratic education.* Princeton, NJ: Princeton University Press.

———. (1997). *Democratic education* (2nd ed.). Princeton, NJ: Princeton University Press.

Gutstein, E. (2006). *Reading and writing the world with mathematics.* New York: Routledge.

Habermas, J. (1968). *Knowledge and human interests.* Boston: Beacon Press.

———. (1981). *The theory of communicative action, Vol. 1, Reason and the rationalization of society.* Boston: Beacon Press.

———. (1984). *The theory of communicative action, Vol. I: Reason and the rationalization of society* (Trans. T. McCarthy). Boston: Beacon Press.

———. (1987). *The theory of communicative action: Vol.11. Life-World and system: A critique of functionalist reason* (Trans. T. McCarthy). Boston: Beacon Press.

———. (1996). The European nation state; its achievements and its limitations: On the past and future of sovereignty and citizenship. *Ratio Juris,* 9(2).

———. (1998). *The inclusion of the other: Studies in political theory.* Cambridge, MA: MIT Press.

———. (1998). What is universal pragmatics? In M. Cooke (Ed.), *On the pragmatics of communication.* Cambridge, MA: MIT Press.

Hahn, S. (2003). *A Nation under our feet: Black political struggles in the rural south from slavery to the great migration.* Cambridge, MA: Belknap Press of Harvard University Press.

Haidara, B. (1990). *Regional programme for the eradication of illiteracy in Africa: Literacy lessons.* Switzerland: International Bureau of Education.

Hammersley, M. (2005). A reply to Korth. In G. Troman, G. Walford, & B. Jeffries (Vol. Eds). *Methodological issues and practices in ethnography. Studies in educational ethnography* (Vol. 11, 168–173). Oxford and London: Elsevier Ltd.

Hann, J. H. (1988). *Apalachee: The land between the rivers.* Gainesville: University of Florida Press.

Harding, S. (1991). *Whose science? Whose knowledge? Thinking from women's lives.* Ithaca, NY: Cornell University Press.

Harding, S. (2003). *The feminist standpoint theory reader: Intellectual and political controversies.* New York: Routledge.

Hegel, G. W. F. (1977). *Phenomenology of spirit* (Trans. A. V. Miller). Oxford: Oxford University Press. (Original work published in 1807).

Heidegger, M. (1971). Building, dwelling, thinking. In M. Heidegger, *Poetry, language, thought* (Trans. A. Hofstadter). New York: Harper and Row.

Hess, F. M. & LoGerfo, L. (2006, May 8). Chicanas from outer space. *National Review Online.* Retrieved January 12, 2007, from http://article.

nationalreview.com/?q=ZDYwOGExMmUxOWY0ZDgxNGQxMGEwZjg4NTNhMzQ2M2M=

Hill, J. P. & Holmbeck, G. N. (1986). Attachment and autonomy during adolescence. *Annals of Child Development,* 3, 145–189.

Hill, W. (1982). *An ethnography of Santa Clara Pueblo New Mexico.* Albuquerque, NM: University of New Mexico Press.

Hirschman, A. (1958). *The strategy of economic development.* CT: Yale University Press.

Hobsbawm, E. (1992). Mass-producing traditions: Europe 1870–1914. In E. Hobsbawm & T. Ranger (Eds.), *The invention of tradition.* Cambridge, MA: Cambridge University Press.

hooks, bell. (2000). *Feminist theory: From margin to center,* 2nd ed. Cambridge, MA: South End Press (Original work published in 1984).

———. (2004).*We real cool: Black men and masculinity.* New York: Taylor and Francis.

Hopwood *v.* Texas, 518 U.S. 1033 (1996).

Horowitz, H. L. (1987). *Campus life: Undergraduate cultures from the end of the 18th century to the present.* New York: Knopf.

Hosmer, D. & Lemeshow, S. (2000). *Applied logistic regression.* New York: John Wiley and Sons.

Hostetler, K. (2005). What is "Good" education research? *Educational Researcher,* 36(6), 16–21.

Huang, Y.-P. (2008). Understanding international graduate instructors: A narrative critical ethnography. Unpublished doctoral dissertation, Indiana University, Bloomington.

Hunter, C. A. (2007). On the outskirts of education: The liminal space of rural teen pregnancy. *Ethnography and Education,* 2(1), 5–92.

Hurtado, S. (1992). The campus racial climate: Contexts for conflict. *The Journal of Higher Education,* 63(5), 539–569.

Hurtado, S. & Carter, D. F. (1997). Effects of college transition and perceptions of the campus racial climate on Latino college students' sense of belonging. *Sociology of Education,* 70, 324–435.

Hurtado, S. & Ponjuan, L. (2005). Latino educational outcomes and the campus climate. *Journal of Hispanic Higher Education,* 4(3), 235–251.

Ingraham, C. (1996). The heterosexual imaginary: Feminist sociology and theories of gender. In S. Seidman (Ed.), *Queer theory/sociology* (169–193). Malden, MA: Blackwell.

Jackson, M. (1998). *Minima ethnographica: Intersubjectivity and the anthropological project.* Chicago, IL: The University of Chicago Press.

Jacoby, R. (2005). *Picture imperfect: Utopian thought for an anti-utopian age.* New York: Columbia University Press.

Jefferson, T. (2002). *Notes on the state of Virginia* (D. Waldstreichen, ed.). New York: Palgrave Macmillan. (Original work published 1785).

Johnson, J. K. (1997). From chiefdom to tribe in northeast Mississippi: The Soto expedition as a window on a culture in transition. In P. Galloway (Ed.),

The Hernando de Soto expedition (295–312). Lincoln: University of Nebraska Press.

Johnson, R. B. and Onwuegbuzie. A. J. (2004). Mixed methods research: A research paradigm whose time has come." *Educational Researcher*, 33(7), 14–26.

Johnson-Bailey, J. (1999). The ties that bind and the shackles that separate: Race, gender, class, and color in a research process. *International Journal of Qualitative Studies in Education*, 12(6), 659–670.

Jones, E. P. (2003). *The known world*. New York: Amistad/Harper Collins.

Jones, J. (1993). The Tuskegee Syphilis experiment: "A moral astigmatism." In S. Harding (Ed.), *The racial economy of science: Toward a democratic future* (275–286). Bloomington, IN: Indiana University Press.

Jules, D. (1991). Building democracy. In M. W. Apple & L. Christian-Smith (Eds.), *The politics of the textbook* (259–287). New York: Routledge.

Kant, I. (1929). *Critique of pure reason* (Trans. N. K. Smith). Boston: Bedford/St. Martin's.

Katie, B. (1998). *Losing the moon: Byron Katie dialogues on non-duality, Truth and Other Illusions*. The Work Foundation, Inc.

Katz, M. B. (2001). *The irony of early school reform: Educational innovation in mid-nineteenth century Massachusetts* (rev. ed.). New York: Teachers College Press. (Original work published 1968).

Keating, A., Ortloff, D. H., & Phillipou, S. (Eds.). (2009) Introduction—Citizenship education curricula: The changes and challenges presented by global and european integration. *Journal of Curriculum Studies*, 2.

Keomea, J. (2004). Dilemmas of an indigenous academic: A Native Hawaiian story. In K. Mutua & B. B. Swadener (Eds.), *Decolonizing research in cross-cultural contexts: Critical personal narratives* (27–44). Albany, NY: State University of New York Press.

Khoi, L. T. (1991). *L'education comparee [Japanese]* (Trans. Y. Maehira). Tokyo: Koji-sha.

Kierkegaard, Søren (1989). *The sickness unto death*. London: Penguin Books.

Kincheloe, J. & McLaren, P. (1998). Rethinking critical theory and qualitative research. In N. Denzin & Y. Lincoln (Eds.), *The landscape of qualitative research: Theories and Issues*. Thousand Oaks, CA: Sage Publications.

———. (2003). Rethinking critical theory and qualitative research. In N. K. Denzin & Y. S. Lincoln (Eds.), *The landscape of qualitative inquiry: Theories and issues* (2nd ed., 433–488). Thousand Oaks, CA: Sage Publications.

Kluckhohn, C. & Leighton, D. (1974). *The Navaho* (Rev. ed.). Cambridge, MA: Harvard University Press (Original work published in 1946).

Korth, B. (2005). Choice, necessity, or narcissism. A feminist does feminist ethnography. In G. Troman et al. (Eds.), *Methodological issues and practices in ethnography. Studies in educational ethnography* (Vol. 11, 131–167). Oxford and London: Elsevier Ltd.

Korth, B. (2006). Establishing universal human rights through war crimes trials and the need for cosmopolitan law in an age of diversity. *Liverpool Law Review*, 27(1), 97–123.

———. (2007). Leaps of faith in social science: Capturing the imaginary in the discourse of the real. *International Journal for Qualitative Methods*, 6(1).

Kuh, G. D. & Whitt, E. J. (1988). The invisible tapestry: Culture in American colleges and universities (ASHE-ERIC, Higher Education Report No. 1). Washington, DC: Association for the Study of Higher Education.

Kuhn, T. S. (1970). *The structure of scientific revolutions*. Chicago: University of Chicago Press.

Ladsen-Billings, G. (2001). *Crossing over to Canaan: The journey of new teachers in diverse classrooms*. San Francisco, CA: Jossey-Bass.

LaFromboise, T. D. & Plake, B. S. (1983). Toward meeting the needs of American Indians. *Harvard Educational Review*, 53(1), 45–51.

Lagemann, E. C. (1997). Contested terrain: A history of education research in the United States, 1890–1990. *Educational Researcher*, 26(9), 5–17.

———. (2000). *An elusive science: The troubling history of education research*. Chicago: University of Chicago Press.

Lamar, C. (1997). Hernando de Soto before Florida: A narrative. In P. Galloway (Ed.), *The Hernando de Soto expedition* (181–206). Lincoln: University of Nebraska Press.

Lather, P. (1991). *Getting smart: Feminist research and pedagogy with/in the postmodern*. New York: Routledge.

———. (1993). Fertile obsession: Validity after poststructuralism. *The Sociological Quarterly*, 34, 673–693.

Lau, V. W. (2007). A particle in motion... Galileo's life. Accessed March 17, 2007 http://math.berkeley.edu/~robin/Galileo/life.html

Lauzon, G. P. (2007). Civic learning through agricultural improvement: Bringing the "loom and the anvil into proximity with the plow" in nineteenth-century Indiana. Unpublished doctoral dissertation, Indiana University, Bloomington, IN.

Law, W. W. (2004). Globalization and citizenship education in Hong Kong and Taiwan. *Comparative Education Review*, 48(3).

Lawrence, A. (2006). Unraveling the White man's burden: A critical microhistory of federal Indian education policy implementation at Santa Clara Pueblo, 1902–1907. Unpublished doctoral dissertation, Indiana University, Bloomington, IN.

Lawrence-Lightfoot, S. (1983). *Good high school: Portraits of character and culture*. New York: Basic Books.

———. (1997). Illumination: Navigating intimacy. In S. Lawrence-Lightfoot J. Davis (Eds.), *The art and science of portraiture* (135–159). San Francisco: Jossey-Bass.

Lawrence-Lightfoot, S. & Davis, J. (1997). *The art and science of portraiture*. San Francisco: Jossey-Bass.

Leon-Portilla, M. (1963). *Aztec thought and culture: A study of the ancient Nahuatl mind* (Trans. J. E. Davis). Norman: University of Oklahoma Press.

Leopold, A. (1968). *A sand county almanac.* Oxford: Oxford University Press. (Original work published 1949).

Lesko, N. (2001). *Act your age: A cultural construction of adolescence.* New York: Routledge Falmer.

Levinson, B. A. U. & Sutton, M. (2001). Introduction. In M. Sutton & B. A. U. Levinson (Eds.), *Policy as practice: Toward a comparative sociocultural analysis of educational policy.* Westport, CT: Ablex Pub.

Levi-Strauss, C. (1966). *The savage mind* (2nd ed.). Chicago: University of Chicago Press.

Lewis, D. L. (1993). *W. E. B. DuBois: Biography of a race, 1868–1919.* New York: Henry Holt.

———. (2000). *W. E. B. DuBois: The fight for equality and the American century.* New York: Henry Holt.

Lincoln, Y. S. & Guba, E. G. (1985). *Naturalistic inquiry.* Thousand Oaks, CA: Sage Publications.

———. (2003). Paradigmatic controversies, contradictions, and emerging confluences. In N. K. Denzin & Y. S. Lincoln (Eds.), *The landscape of qualitative inquiry: Theories and issues* (2nd ed., 245–252). Thousand Oaks, CA: Sage Publications.

Linde, C. (1993). *Life stories: The creation of coherence.* New York: Oxford University Press.

Lipka, J., Mohatt, G., & the Ciulistet Group. (1998). *Transforming the culture of schools: Yup'ik Eskimo examples.* Mahwah, NJ: Lawrence Erlbaum.

Livingston, G. (2003). Chronic silencing and struggling without witness: Race, education and the production of political knowledge. Unpublished dissertation, University of Wisconsin, Madison.

Llagas, C. & Snyder, T. D. (2003). *Status and trends in the education of Hispanics.* Washington, DC: U.S. Department of Education, National Center for Educational Statistics. (NCES 2003–2008).

Locks, A. M., Hurtado, S., Bowman, N. A., & Oseguera, L. (2008). Extending notions of campus climate and diversity to students' transition to college. *The Review of Higher Education,* 31(3), 257–286.

Lomawaima, K. T. (2000). Tribal sovereigns: Reframing research in American Indian education. *Harvard Educational Review,* 70(1), 1–21.

Lomawaima, K. T. & McCarty, T. L. (2006). *To remain an Indian: Lessons in democracy from a century of Native American education.* New York: Teachers College Press.

Long, S. (1997). *Regression models for categorical and limited dependent variables.* Thousand Oaks, CA: Sage Publications.

Lopez, B. (1990). *The rediscovery of North America.* New York: Random House.

Luke, D. (2004). *Multilevel modeling.* Thousand Oaks, CA: Sage Publications.

Malone, S. (2003). Ethics at home: Informed consent in your own backyard. *International Journal of Qualitative Studies in Education, 16(6)*, 797–815.

Mann, C. C. (2005). *1491: New revelations of the Americas before Columbus.* New York: Alfred A. Knopf.

Mannheim, K. (1936). *Ideology and utopia.* New York: Harvest Books.

Marshall, C. (1999). Researching the margins: Feminist critical policy analysis. *Educational Policy*, 13(1), 59–76.

Martin, B. (1992). Scientific fraud and the power structure of science. *Prometheus*, 10(1), 83–98.

———. (2000). Directions for liberation science. *Philosophy and Social Action*, 26(1–2), 9–21.

Martín-Baró, I. (1994). *Writings for a liberation psychology.* Cambridge, MA: Harvard.

Marx, Leo. (1964). *Machine in the garden.* Oxford, UK: Oxford University Press.

Marx, K. (1999). *Das capital.* Washington, DC: Gateway Editions.

Maxwell, J. A. (1992). Understanding and validity in qualitative research. *Harvard Educational Review*, 62, 279–300.

———. (2002). Understanding and validity in qualitative research. In A. M. Huberman & M.B. Miles (Eds.), *A Qualitative Researcher's Companion.* Thousand Oaks, CA: Sage Publications, 37–64.

May, T. (2004). *The moral theory of poststructuralism.* University Park, PA: Pennsylvania State University Press.

Mays, N. and Pope, C. (2000). Qualitative research in health care: Assessing quality in qualitative research. *British Medical Journal*, 320, 50–52.

McCarn, S. & Fassinger, R. (1996). Revisioning sexual minority identity formation: A new model of lesbian identity and its implications for counseling and research. *The Counseling Psychologist*, 24(3).

McCarty, T. L. (2002). *A place to be Navajo: Rough rock and the struggle for self-determination in indigenous schooling.* Mahwah, NJ: Lawrence Erlbaum.

McCowan, S. (1903). Letter to C. J. Crandall regarding the participation of Pueblo Indians in the 1904 St. Louis Expo, May 14, 1903. *Records of the Bureau of Indian Affairs, Northern Pueblos Agency, Miscellaneous Correspondence and Reports, 1868–1904.*

McGrath, M. (2007). *The long exile: A tale of inuit betrayal and survival in the high Arctic.* New York: Alfred A. Knopf.

McLaren, P. L. & Giarelli, J. M. (Eds.). (1995). *Critical theory and educational research.* Albany, NY: State University of New York Press.

———. (1934). *Mind, self, and society: Explorations by a social behaviorist* (C. W. Morris, Ed.). Chicago: University of Chicago Press.

Meier, G. M. and Stiglitz, J. E. (Eds.). (2000). *Frontiers of development economics: The future in perspective.* Washington, DC: The World Bank.

Menard, S. (2002). *Applied logistic regression analysis* (2nd ed.). Thousand Oaks, CA: Sage Publications.

Merleau-Ponty, M. (1962). *Phenomenology of perception*. New York: Humanities Press.

Merriam, S., Johnson-Bailey, J., Lee, M., Kee, Y., Ntseane, Gabo, & Muhamad, M. (2001). Power and positionality: Negotiating insider/outsider status within and across Cultures. *International Journal of Lifelong Education*, 20, 5.

Meyer, R. W. (1977). *The village Indians of the Upper Missouri: The Mandans, Hidatsas, and Arikaras*. Lincoln: University of Nebraska Press.

Mihesuah, D. (1998a). *Natives and Academics: Researching and writing about American Indians*. Lincoln, NE: University of Nebraska Press.

———. (1998b). Introduction. In D. A. Mihesuah (Ed.), *Natives and academics: Researching and writing about American Indians*. Lincoln: University of Nebraska Press.

———.(2000). American Indians, anthropologists, pothunters, and repatriation: Ethical, religious, and political differences. In D. Mihesuah (Ed.), *Repatriation reader: Who owns indian remains?* (95–105). Lincoln, NE: University of Nebraska Press.

Mihesuah, D. A. and Wilson, A. C. (Eds.). (2004). *Indigenizing the academy: Transforming scholarship and empowering communities*. Lincoln, NE: University of Nebraska Press.

Milanich, J. T. (1994). *Archaeology of precolumbian Florida*. Gainesville: University Press of Florida.

Miller, J. J. (1998). Foreword. In C. R. Ewen & J. J. Hann (Eds.), *Hernando de Soto among the Apalachee: The archaeology of the first winter encampment* (x–xi). Gainesville: University Press of Florida.

Mills, C. (1997). *The racial contract*. Ithaca, NY: Cornell University Press.

Minh-ha, T. (1989). *Women, native, and other: Writing post coloniality and feminism*. Bloomington, IN: Indiana University Press.

Monaghan, E. J. (2005). *Learning to read and write in colonial America*. Amherst: University of Massachusetts Press.

Moustakas, C. (1994). *Phenomenological research methods*. Thousand Oaks, CA: Sage Publications.

Muir, E. (1991). Introduction: Observing trifles. In *Microhistory and the lost peoples of Europe* (vii–xxviii). Baltimore and London: Johns Hopkins University Press.

Muller, A. & Murtagh, T. (2002). Literacy: The 877 million left behind. *Education Today*, 2, 4–7.

Murguía, E., Padilla, R. V., & Pavel, D. M. (1991). Ethnicity and the concept of social integration in Tinto's model of institutional departure. *Journal of College Student Development*, 32(5), 433–439.

Nabhan, G. P. (1997). *Cultures of habitat: On nature, culture, and story*. Washington, DC: Counterpoint.

Nabhan, G. P. and Trimble, S. (1994). *The geography of childhood: Why children need wild places*. Boston: Beacon Press.

National Commission on Excellence in Education. (1983). *A nation at risk.* Retrieved March 26, 2007, from http://www.ed.gov/pubs/NatAtRisk/index.html.

National Museum of the American Indian. (2007). *Do all Indians live in tipis?: Questions and answers from the National Museum of the American Indian* (1st ed.). New York, NY: Harper Collins Publishers.

Newsom, C. N., Ridenour, C., & Kinnucan-Welsh, K. (2001). Is the tape off? African Americans' spontaneous discussions of race and racism when the researcher is also African American. Paper presented at the Annual Meeting of the American Educational Research Association, Seattle, WA.

Newton, E. (2000). *Margaret mead made me gay: Personal essays, Public Ideas.* United States: Duke University Press.

Nichols, R. L. (2003). *American Indians in U.S. history.* Norman: University of Oklahoma Press.

Nisargadatta, S. M. (1988). *I am that: Talks with Sri Nisargadatta Maharaj* (Ed. S. S. Dikshit; Trans. M. Frydman). Durham, NC: Acorn Press.

Nora, A. (1987). Determinants of retention among Chicano college students: A structural model. *Research in Higher Education, 26,* 31–59.

———. (2001). *How minority students finance their higher education.* New York: ERIC Clearinghouse on Urban Education, Office of Educational Research and Improvement, U.S. Department of Education. (ERIC Document Reproduction Service NO. ED460243).

———. (2002). A theoretical and practical view of student adjustment and academic achievement. In W. Tierney & L. Hagedorn (Eds.), *Increasing access to college: Extending possibilities for all students.* Albany, NY: State University of New York Press.

———. (2004a). The role of *habitus* and cultural capital in choosing a college, transitioning from high school to higher education, and persisting in college among minority and non-minority students. *Journal of Hispanic Higher Education, 3*(2), 180–208.

———. (2004b). Access to higher education for Hispanic students: Real or illusory? In J. Castellanos & L. Jones (Eds.), *The majority in the minority.* Sterling, VA: Stylus.

———. The future of research on Hispanic students. In A. Gloria (Ed.), Sterling, WV: Stylus Publishing, LLC.

Nora, A., Barlow, L., & Crisp, G. (2005). An assessment of hispanic students in four-year institutions of higher education. In J. Castellanos, A. M. Gloria & M. Kamimura (Eds.), *The Latina/o pathway to the Ph.D.* (55–77). Sterling, WV: Stylus Publishing, LLC.

Nora, A. and Cabrera, A. F. (1993). The construct validity of institutional commitment: A confirmatory factor analysis. *Research in Higher Education, 34*(2), 243–262.

———. (1996). The role of perceptions of prejudice and discrimination on the adjustment of minority students to college. *Journal of Higher Education, 67*(2), 120–148.

Nora, A., Cabrera, A. F., Hagedorn, L. S., & Pascarella, E. T. (1996). Differential impacts of academic and social experiences on college-related behavioral outcomes across different ethnic and gender groups at four-year institutions. *Research in Higher Education*, 37(4), 427–451.

Nora, A. and Crisp, G. Hispanics and Higher Education: An Overview of Research, Theory and Practice. In J. Smart and S. Thomas (Eds.), *Handbook of Higher Education: Research and Theory*.

————. (in press). Mentoring students: Conceptualizing and validating the multi-dimensions of a support system. *Journal of College Student Retention: Theory and Practice*, 9(1).

Nora, A. & Garcia, V. (April, 1999). Attitudes related to remediation among developmental students in higher education. Paper presented at the annual meeting of the American Educational Research Association.

Nurkse, R. (1953). *Problems of capital formation in underdeveloped countries*. Oxford: Basil Blackwell.

Odora Hoppers, C. A. (Ed.). (2002). *Indigenous knowledge and the integration of knowledge systems: Towards a philosophy of articulation*. Claremont, South Africa: New Africa Books.

Oleson, V. L. (2003). Feminisms and qualitative research at and into the millennium. In N. K. Denzin & Y. S. Lincoln (Eds.), *The landscape of qualitative inquiry: Theories and issues* (2nd ed., 332–432). Thousand Oaks, CA: Sage Publications.

Olivas, M. A. (2005). Higher education as "Place": Location, race, and college attendance policies. *The Review of Higher Education*, 28(2), 169–189.

Oliver, A., Tavares, H., Gardin, L. A., Apple, M. W., Cho, M. K., Aasen, P. et al. (2003). *The state and the politics of knowledge*. New York: Routledge.

Olson, S. F. (2001). *The meaning of wilderness: Essential articles and speeches* (ed. D. Backes). Minneapolis, MN: University of Minnesota Press.

Ortloff, D. H. (2006). Becoming European: A framing analysis of three countries' civics education curricula. *European Education*, 37(4), 35–49.

Ortloff, D. H. & Frey, C. J. (November, 2007). Blood relatives?: Language, immigration and education of Ethnic Returnees in Germany and Japan." *Comparative Education Review*, 51(4).

Ostler, J. (2004). *The Plains Sioux and U.S. colonialism from Lewis and Clark to Wounded Knee*. New York: Cambridge University Press.

Oxford English Dictionary, 2. E. (1989). Education. Retrieved September 6, 2008, from http://dictionary.oed.com/cgi/entry/50072205?single=1&query_type=word&queryword=education&first=1&max_to_show=10.

Padilla, R. V., Trevino, J., Gonzalez, K., & Trevino, J. (1997). Developing local models of minority student success in college. *Journal of College Student Development*, 38, 125–135.

Pampel, F. (2000). *Logistic regression: A primer*. Thousand Oaks, CA: Sage Publications.

Parker, I. (1997). Discursive psychology. In. D. Fox & I. Prilleltensky (Eds.), *Critical psychology: An introduction* (284–298). Thousand Oaks, CA: Sage Publications.

Parks, D. R. (Ed.). (1991). *Traditional narratives of the Arikara Indians* (Vols. 1–4). Lincoln: University of Nebraska Press.

Pascarella, E. T. & Terenzini, P. T. (2005). *How college affects students: A third decade of research*. San Francisco: Jossey-Bass.

Payne, E. (January 6, 2005). Adolescent females self-labeling as lesbian and the gender binary: A critical life story study. Paper presented at the Qualitative Inquiry Program Area Conference, Athens, GA.

Pedhazur, E. J. (1982). *Multiple regression in behavioral research* (2nd Ed.). New York: Holt, Rinehart, and Winston.

Penrose, R. (1994). *Shadows of the mind: A search for the missing science of consciousness*. New York: Oxford University Press.

Pincus, J. (Ed.). (1980). *Educational evaluation and the public policy setting* (R-2505-RC) Santa Monica: Rand Corporation.

Plummer, K. (1996). Symbolic interactionism and the forms of heterosexuality. In Seidman (Ed.), *queer theory/sociology*. Malden, MA: Blackwell.

Platt, J. & Weber, H. (1984). Speech convergence miscarried: An investigation into inappropriate accommodation strategies. *International Journal of the Sociology of Language*, 46, 131–146.

Pope, J. H., Aufderheide, T. P., Ruthazer, R., Woolard, R. H., Feldman, J. A., Beshansky, J. R. et al. (2000). Missed diagnoses of acute cardiac ischemia in the emergency department. *New England Journal of Medicine*, 342(16), 1163–1170.

Raizen, S. & Rossi, P. (Eds.). (1981). *Program evaluation in education: When? How? To what ends?* Washington, DC: National Academy of Sciences.

Raudenbush, S. & Bryk, A. (2002). *Hierarchical linear models: Applications and data analysis methods* (2nd ed.). Thousand Oaks, CA: Sage Publications.

Raudenbush, S., Bryk, A., Cheong, Y. F., Congdon, R., & Toit, M. D. (2004). *HLM6: Hierarchical linear & nonlinear modeling*. IL: Scientific Software International.

Rendon, L. I. (1994). Validating culturally diverse students: Toward a new model of learning and student development. *Innovative Higher Education*, 19(1), 23–32.

Rendon, L. I., Jalomo, R., & Nora, A. (2000). Minority student persistence. In J. Braxton (Ed.), *Rethinking the departure puzzle: New theory and research on college student retention*. Nashville, Vanderbilt University Press.

Rendon, L. I., Novack, V., & Dowell, D. (2005). Testing race-neutral admissions models: Lessons from California State University-Long Beach. *The Review of Higher Education*, 28(2), 221–244.

Reséndez, A. (2007). *A land so strange: The epic journey of Cabeza de Vaca*. New York: Basic Books.

Revel, J. (1995). The critique of social history. In *Histories: French constructions of the past* (Vol. 1, 492–502). New York: New Press.

Reyhner, J. & Eder, J. (2004). *American Indian education: A history*. Norman: University of Oklahoma Press.

Robson, C. (2002). *Real world research*. Oxford: Blackwell Publishers.

Rosaldo, R. (1989). *Culture and truth: The remaking of social analysis*. Boston: Beacon.

Rostow, W. W. (1960). *The stages of economic growth*. New York: Cambridge University Press.

Rury, J. L. (2005). *Education and social change: Themes in the history of American schooling* (2nd ed.). Mahwah, NJ: Lawrence Erlbaum Associates, Publishers.

Samoff, J. (1990). "Modernizing" a socialist vision: Education in Tanzania. In M. Carnoy & J. Samoff (Eds.), *Education and social transition in the third world* (209–273). Princeton, NJ: Princeton University Press.

Sarris, G. (1993). *Keeping slug woman alive: A holistic approach to American Indian texts* (214). Berkeley: University of California Press.

Schama, S. (1995). *Landscape and memory*. New York: Alfred A. Knopf, INC.

Schissler, H. & Soysal, Y. N. (2005). *The nation, Europe and the World: Textbooks and curricula in transition*. New York: Beghahn Books.

Schultz, T. (1971). *Investment in human capital*. New York: The Free Press.

Sen, A. (1997). Editorial: Human capital and human capability. *World Development*, 25(12), 1959–1961.

———. (1999). *Commodities and capabilities* (Oxford India Paperbacks ed.). New Delhi: Oxford University Press.

———. (2000). *Development as freedom*. New York: Anchor Books.

———. (2002). Capability and well-being. In M. Nussbaum & A. Sen (Eds.), *The quality of life*. Oxford: Oxford University Press.

Setran, D. P. (2007). *The college "Y": Student religion in the era of secularization*. New York: Palgrave/Macmillan.

Sevier, A., Clifford, N., & de la Rosa, L. (February 2, 1848). Treaty of Guadalupe Hidalgo. Retrieved from http://www.azteca.net/aztec/guadhida.html.

Shadish, W. R., Cook, T. D., and Campbell, D. T. (2002). *Experimental and quasi-experimental designs for generalized causal inference*. Boston, MA: Houghton Mifflin Company.

Shaffer, L. N. (1992). *Native Americans before 1492: The mound-building centers of the eastern woodlands*. Armouk, NY: M. E. Sharpe.

Shannon, T. J. (2000). *Indians and colonists at the crossroads of empire: The Albany congress of 1754*. Ithaca, NY: Cornell University Press.

Shavelson, R. J. & Towne, L. (2002). *Scientific research in education*. Washington, DC: National Academy Press.

Slaughter, S. (1988). National higher education policies in a global economy. In J. Currie & J. Newson (Eds.), *Universities and Globalization: Critical Perspectives*. Thousand Oaks, CA: Sage Publications.

Slocum, W. F. (1904). The world's fair as an educative force. *Outlook (1893–1924)*, 785.

Smith, F. T. (1996). *The Caddos, the Wichitas, and the United States, 1846–1901*. College Station: Texas A and M Press.

Smith, J. S. (2008). Expectations for research intensify alongside accountability. *Indiana Insight*, Spring Issue, 20–23.

Smith, L. T. (1999). *Decolonizing methodologies: Research and indigenous peoples.* London: Zed Books Ltd.

Snijders, T. & Bosker, R. (2003). *Multilevel analysis: An introduction to basic and advanced multilevel modeling.* Thousand Oaks, CA: Sage Publications.

Snyder, Gary. (1995). *From discovering a sense of place.* Portland, OR: Northwest Earth Institute.

Soysal, Y. N., Bertilotti, T., & Mannitz, S. (2005). Projections of identity in French and German history and civics textbooks. In H. Schissler & Y. N. Soysal (Eds.), *The nation, Europe and the World: Textbooks and Curricula in Transition.* New York: Berghahn Books.

Spivak, G. (1988). Can the subaltern speak? In C. Nelson & L. Grossberg (Eds.), *Marxism and the interpretation of culture* (271–313). Urbana: University of Illinois Press.

Spradley, J. (1979). *The ethnographic interview.* New York: Holt Rinehart and Winston.

Spring, J. (2008). *The American school: From the Puritans to No Child Left Behind* (7th ed.). New York: McGraw-Hill.

St. John, E. P., Cabrera, A. F., Nora, A., & Asker, E. H. (2001). Economic perspectives on student persistence. In J. Braxton's (Ed.), *Rethinking the departure puzzle: New theory and research on college student retention.* Nashville, TN: Vanderbilt University Press.

Stegner, Wallace. (1995). *Discovering a sense of place.* Portland, OR: The Northwest Earth Institute.

Steigman, J. D. (2005). *"La Florida del Inca" and the struggle for social equality in colonial spanish America.* Tuscaloosa: University of Alabama Press.

Stoller, P. (1997). *Sensuous scholarship.* Philadelphia, PA: University of Pennsylvania Press.

Storr, R. (1961). The education of history: Some impressions. *Harvard Educational Review*, 31, 124–135.

Storr, R. J. (1976). The role of education in American history: A memorandum. *Harvard Educational Review*, 46(3), 331–354.

Streeten, P. (1981). *First things first: Basic human needs in developing countries.* Washington, DC: World Bank Publication.

Streibel, M. J. (1998). The importance of physical place and lived topographies. Paper presented at the National Convention of the Association for Educational Communications and Technology, St. Louis, MO.

Sutton, M. (2005). The globalization of multicultural education. *Indiana Journal of International Legal Studies*, 12(1), 97–108.

Svenningsen, R. (1980). Preliminary inventory of the Pueblo records created by field offices of the Bureau of Indian Affairs, record group 75. General Services Administration, Washington, DC.

Swisher, K. G. (1998). Why Indian people should be the ones to write about Indian education. In D. A. Mihesuah (Ed.), *Natives and academics: Researching and writing about American Indians.* Lincoln: University of Nebraska Press.

Szasz, M. C. (1988). *Indian Education in the American Colonies.* Albuquerque: University of New Mexico Press.

Tajfel, H. (Ed.). (1978). *Differentiation between social groups.* London: Academic Press.

Tashakkori, A. & Teddlie, C. (Eds.). (2003). *Handbook of mixed methods in social and behavioral research.* Thousand Oaks, CA: Sage Publications.

Teitelbaum, K. (1993). *Schooling for good rebels.* Philadelphia: Temple University Press.

The American College of Physicians. (2000). Primer on probability and odds and interpreting their ratios. *Effective Clinical Practice, 3*(3), 145–146.

The Campbell Collaboration (2000). Available at http://campbellcollaboration.org.

The No Child Left Behind Act (NCLB,2001). Available at http://www.ed.gov/nclb/landing.jhtml?src=pb (Accessed February 21, 2009).

The St. Louis Fair. (1903). *Outlook (1893–1924), 73*(17), 952–953.

The What Works Clearinghouse (2002). Available: http://ies.ed.gov/ncee/wwc (Accessed February 21, 2009).

The World Bank. (1993). Human resource development survey: Individual/household questionnaire: The World Bank.

————. (1997). Tanzania human resource development survey. Retrieved November 1, 2004, from http://www.worldbank.org/lsms/country/tza/tanzhome.html.

The World Bank and University of Dar es Salaam. (1993). *Human resource development survey: Interviewer's manual:* The World Bank.

Theunissen, M. (2005). *Kierkegaard's concept of despair.* Oxford and Princeton: Princeton University Press.

Thorne, B. & Luria, Z. Sexuality in children's daily worlds. In C. Williams & A. Stein (Eds.), *Sexuality and gender* (127–141). Malden, MA: Blackwell Publishing.

Tierney, W. G. (1992). An anthropological analysis of student participation in college. *Journal of Higher Education, 62*(6), 603–618.

Tosh, J. (2002). *The pursuit of history: Aims, methods and new directions in the study of modern history.* London: Longman.

Trennert, R. A., Jr. (1987). Selling Indian education at world's fairs and expositions, 1893–1904. *American Indian Quarterly, 11*(3), 203–220.

Trigger, B. G. & Washburn, W. E. (Eds.). (1996). *The Cambridge history of the Native peoples of the Americas* (Vol. 1). New York: Cambridge University Press.

True, C. (1904a). Letter to superintendent concerning a Santa Clara Pueblo woman attending the St. Louis exposition to demonstrate her pottery, January 20, 1904. *Letters Received from day school teacher Clara D. True, 08/29/1902–11/30/1906.*

————. (1904b). Letter to superintendent concerning additional women potters to go to St. Louis exposition, February 18, 1904. *Letters received from day school teacher Clara D. True, 08/29/1902-11/30/1906.*

True, C. (1904c). Letter to superintendent pertains to the woman potter and Her husband and their trip to the St. Louis exposition, March 31, 1904. *Letters received from day school teacher Clara D. True, 08/29/1902–11/30/1906.*

———. (1904d). Letter to superintendent requesting permission to allow Pedro Chiquito Cajete and his daughter to go to the St. Louis exposition, April 14, 1904. *Letters Received from day school teacher Clara D. True, 08/29/1902–11/30/1906.*

———. (1904e). Letter to superintendent concerning the Santa Clara Indians at the St. Louis exposition, May 17, 1904. *Letters received from day school teacher Clara D. True, 08/29/1902–11/30/1906.*

———. (1909). The eexperiences of a woman Indian agent. *Outlook (1893–1924)*, 331.

Tugendhat, E. (1986). *Self-consciousness and Self-determination.* Cambridge, MA: The MIT Press.

Turner, V. (1986). Dewey, Dilthey, and drama: An essay in the anthropology of experience. In *The anthropology of experience* (33–44). Urbana and Chicago: University of Illinois Press.

UNESCO. (2005). *International literacy day 2005 to focus on sustainable development.* Retrieved September 10, 2005, from http://portal.unesco.org/education/en

UNICEF. (1998). *The state of the world's children 1999.* NY: UNICEF.

———. (2007). *The state of the world's children 2008.* NY: UNICEF.

United States of America *v.* José Juan Lucero. (1869). 1 U.S. 422.

United States *v.* Sandoval, 231 U.S. 28, 34 S.Ct. 1, 58 L.Ed. 107 (District Court of the United States for the District of New Mexico 1913).

Urban, W. J. & Wagoner, J. L., Jr. (2004). *American education: A history* (3rd ed.). New York: McGraw-Hill.

Van Maanen, J. & Barley, S. R. (1985). Occupational communities: Culture and control in organizations. *Research in organizational behavior*, 6, 287–365.

Venezia, A., Kirst, M. W., & Antonio, A. (2003). *Betraying the college dream: How disconnected K-12 and postsecondary education systems undermine student aspirations.* Stanford, CT: The Bridge Project.

Vidich, A. J. & Lyman, S. (2003). Qualitative methods: Their history in sociology and anthropology. In N. K. Denzin & Y. S. Lincoln (Eds.), *The landscape of qualitative inquiry: Theories and issues* (2nd ed., 55–129). Thousand Oaks, CA: Sage Publications.

Wagner, D. A. (1990). Literacy assessment in the third world: An overview and proposed schema for survey use. *Comparative Education Review*, 34(1), 112–138.

———. (1992). World literacy: Research and policy in the EFA decade. *Annals of the American Academy of Political and Social Science*, 520 (World Literacy in the Year 2000), 12–26.

Walat, M. (2006). Towards an intercultural frame of mind: Citizenship in Poland. In G. Alred, M. Byram, & M. Fleming (Eds.), *Education for intercultural citizenship.* Clevedon: Multilingual Matters.

Wallace, A. F. C. (1999). *Jefferson and the Indians: The tragic fate of the first Americans.* Cambridge, MA: Belknap Press of Harvard University Press.

Warren, D. (2005a). Slavery as an American educational institution: Historiographical inquiries. *Journal of Thought,* 40(4), 41–54.

———.(2005b). The Wonderful Worlds of the Education of History. *American Educational History Journal,* 32(1), 108–115.

Warren, J. R. (1968). Student perceptions of college subcultures. *American Educational Research Journal,* 5, 213–232.

Weber, M. (1949). Objectivity in social science and social policy. In E. Shils and H. Finch (Eds. and Trans.), *The Methodology of the Social Sciences* (49–112). New York: The Free Press (Original work published in 1904).

Weber, M. (1968). *Economy and society* (Trans. G. Roth and C. Wittich). New York: Bedminster Press (Original work published in 1921).

Weber, R. P. (1985). *Basic content analysis.* Newbury Park, CA and London: Sage Publications.

Weinberg, S. (2001). *Facing up: Science and its cultural adversaries.* Cambridge, MA: Harvard University Press.

Westbury, I., Hopmann, S., & Riquarts, K. (2000). *Teaching as a reflective practice: The german didaktik tradition.* Mahwah, New Jersey: Lawrence Erlbaum Associates.

Whitty, G. (1974). Sociology and the problem of radical educational change. In M. Flude & J. Ahier (Eds.), *Educability, schools, and ideology* (112–137). London: Halstead Press.

Wiesel, E. (1986). Acceptance speech for the Noble Prize. Retrieved October 17, 2008, from http://www.eliewieselfoundation.org/nobelprizespeech.aspx

Wildavsky, A. (1979). *Speaking truth to power.* Boston: Little Brown.

Wilkins, M. (1999, May). Social responsibility in science. Paper presented at the Institute of Science in Society, London.

Williams, C. & Stein, A. (Eds.). (2002). *Sexuality and gender.* Malden, MA: Blackwell Publishing.

Williams, R. (1961). *The long revolution.* London: Chatto and Windus.

———. (1977). *Marxism and literature.* New York: Oxford University Press.

Wilson, A. C. (1998). Grandmother to Granddaughter: Generations of oral history in a Dakota family. In Mihesuah, D. (Ed.), *Natives and academics: Researching and writing about American Indians* (27–36). Lincoln, NE: University of Nebraska Press.

Wilson, A. C. (2004). Reclaiming our humanity: Decolonizing and the recovery of indigenous knowledge. In D. A. Mihesuah & A.C. Wilson (Eds.), *Indigenizing the academy: Transforming scholarship and empowering communities* (69–87). Lincoln, NE: University of Nebraska Press.

Wilson, J. (1998). *The earth shall weep: A history of Native America.* New York: Grove Press.

Winkle-Wagner, R. (2006). The unchosen me: Institutionally imposed identity and women's college experiences. Unpublished doctoral dissertation, Indiana University, Bloomington.

Winkle-Wagner, R. (In Press). *The unchosen me: The creation of race and gender in college.* Baltimore, MD: The Johns Hopkins University Press.

Wittgenstein, L. (1963). *Philosophical investigations.* Oxford: Blackwell.

Wong, T. H. (2002). *Hegemonies compared.* New York: Routledge.

Woodhouse, E., Hess, D., Breyman, S., & Martin, B. (2002). Science studies and activism: Possibilities and problems for reconstructivist agendas. *Social Studies of Science,* 32(2), 297–319.

Wright, E. O. (1985). *Classes.* New York: Verso.

Yazzie-Mintz, T. (2000, May). Holding a mirror to "Eyes wide shut": The role of Native cultures and languages in the education of American Indian students. Paper presented at the American Indian and Alaska Native Education Research Agenda Conference, Albuquerque, NM.

———. (2002). Culture deep within us: Culturally appropriate curriculum and pedagogy in three Navajo teachers' work. Unpublished doctoral dissertation, Harvard University, Cambridge, MA.

———. (2006). Weaving circles of inquiry: Non-linear nomadic expressions of my research identities. Paper presented at the Annual meeting of the American Anthropological Association, Council on Anthropology and Education, San Jose, CA. November 15–19, 2006.

Youdell, D. (in process). *After identity: Power and politics in education.* London: Routledge.

Young, R. (2003). *Postcolonialism.* New York: Oxford University Press.

Zinn, M. B. (1979). Field research in minority communities: Ethical, methodological and political observations by An Insider. *Social Problems,* 27(2), 209–219.

Index

Printed and bound in Great Britain by
CPI Antony Rowe, Chippenham and Eastbourne